MW01030765

Light Water

Light Water

How the Nuclear Dream Dissolved

Irvin C. Bupp & Jean-Claude Derian

Basic Books, Inc., Publishers New York

All quotations that originally appeared in French have been translated into English by Jean-Claude Derian.

Library of Congress Cataloging in Publication Data

Bupp, Irvin C.
Light water.

Includes index.
1. Atomic power industry. 2. Water-cooled reactors. 3. Public relations—Atomic energy industries. I. Derian, Jean-Claude, joint author. II. Title.
HD9698.A2B78 338.4'7'621483 77-20419
ISBN: 0-465-04107-8

Printed in the United States of America
DESIGNED BY VINCENT TORRE
10 9 8 7 6 5 4 3 2 1

In the greatest part of our concernment God has afforded us only the twilight, as I may so say, of probability, suitable, I presume to that state of mediocrity and probationership He has been pleased to place us in here.

JOHN LOCKE
"Essay Concerning
Human Understanding"

CONTENTS

Part III

THE TWILIGHT OF PROBABILITY

PREFACE

FRIENDS continually ask us if we are "for" or "against" nuclear power. The apparent simplicity of the choice immobilizes us. The only honest, simple response we can offer is a wishy-washy "Yes, but . . . ." Understandably, this is easily misinterpreted as a code for "against." Let us, therefore, stress at the outset that we are "for" nuclear power in the sense that we share with the majority of our colleagues and associates the belief that it has an important job to do. The world needs nuclear power in the immediate future to meet an important share of energy requirements.

Both of us have spent about fifteen years in close professional association with the nuclear power programs of our respective countries, France and the United States. Both of us have been employees of our countries' Atomic Energy Commissions in capacities that provided a perspective on important policy-level deliberations and actions. Both of us number many colleagues, acquaintances, and close friends among our governments' nuclear power establishment. We have assisted and advised government officials with large policy-making responsibilities in atomic energy and related areas. Finally, as teachers and researchers, we both have participated in numerous, non-governmental studies and reports on atomic energy matters. It was through one such study that we first met in 1973 in Cambridge and discovered the similarity of our backgrounds and experiences. Even the one significant difference—Derian's main academic training and degrees are in physics and engineering; Bupp's in political science and economics—argued for collaboration.

Jean-Claude Derian first formulated the question which was to define the scope of our initial research: How had American light water reactor technology come to dominate the world-wide market for nuclear power plants so thoroughly? This was during the winter of 1973-1974; a time when most informed persons—including ourselves—believed that electricity from nuclear power would be the dominant new energy source in the coming ten to fifteen years, and probably for the rest of the century as well.

The answer to Derian's question began to emerge from our subsequent research over the next year. (You will find the story in the first five chapters of this book.) In the course of our research we found accumulating evidence for something we had originally sensed but had not clearly articulated: light water technology's economic success was linked directly to the spreading controversy over nuclear power.

Our interpretation of how light water became the dominant nuclear power technology and of how this process seems to be linked to the reactor safety controversy will not please those who are committed to atomic energy's role as a vital part of our future. Obviously, and especially in the "social sciences," no research can be exhaustive; no conclusion unambiguously supported by all available evidence; and no judgment wholly free from preconception or bias. Obviously, there are parts of the massive and complex nuclear power story about which we are either essentially ignorant or, at best, weakly informed. For example, we have had no access to the internal records—financial or otherwise—of the principal reactor manufacturers.

But, we believe that our interpretation of the salient events and decisions of the past twenty-five years is a responsible and accurate one within the inevitable limitations of time, resources, and human memory. We also believe it to be a fair one. The story we piece together has never been conceived as an exposé, much less as yet another weapon for the use of one side or the other of the nuclear power controversy.

Rather, we believe that a hard look at what appears to have happened in the past is a key to a more acceptable future.

Among all of the comments we have received on our manuscript during the past year, the one which has struck us most sharply was from a French government official. "What you have told," he said, "is the story of the abuse of a technology." We have appropriated the phrase as the title of one of the concluding chapters, and hereby, acknowledge this debt.

Naturally, we have accumulated many other debts, which it is our customary duty to acknowledge while at the same time absolving all of our numerous creditors from responsibility for content.

Therefore, let us first acknowledge our immense debt to those who made our three-year collaboration possible. Our research has been jointly sponsored by the Center for Policy Alternatives, Massachusetts Institute of Technology, and the Division of Research of the Graduate School of Business Administration, Harvard University. We have also received support from the "Management Beyond Business" Program of Harvard's Graduate School of Business Administration and the "Program for Science and International Affairs," Harvard University. Of course, institutions provide the money, people lend the spirit. Two persons have given us such invaluable help, encouragement, and advice: Professor J. Herbert Hollomon, Director of the Center for Policy Alternatives, and Professor Robert B. Stobaugh, Director of the Business School Energy Research Project. Our debt to these two men is too large to describe.

For research assistance and intellectual stimulation during the early stages of our work, we are deeply grateful to Marie-Paul Donsimoni and Robert Treitel. The advice and encouragement which we received from Pierre Aigrain, former Délégué Général à la Recherche Scientifique et Technique in France, during his stay in the U.S. as M.I.T. Visiting Professor, were essential to the initial orientation of our research. Without Alan Jakimo's intelligence, sense of responsibility and unfailing good humor in the face of outrageous demands for the past several months, this manuscript probably would not have been completed in the present decade. At a crucial stage in our writing Rab Bertelson revealed to us the difference between literate prose and academic/bureaucratic jargon. The book's all too evident remaining failings in this area are entirely our own.

Each of us has a number of colleagues, friends, and former teachers who in a variety of ways, often indirect but nonetheless important, have contributed to our joint endeavor. In the United States these include Richard G. Hewlett, William C. Kriegsman, George C. Lodge, Woodford B. McCool, John G. Palfrey, Don K. Price, and Milton Shaw; in France: Etienne Bauer, Achille Ferrari, Bernard Laponche, Jean-Marie Martin, and Louis Puiseux. Naturally these acknowledgments should not be taken to imply that all or even any of these individuals agree with the substance of our conclusions and interpretations.

The unusually tiresome, and at times apparently endless, typing and clerical tasks which accompany any effort of this sort have been shared, always with competence and good cheer, among: Rose Giacobbe, Emily Hood, and Rona McCrensky.

Cambridge, Massachusetts
Paris, France
January 1978

Light Water

INTRODUCTION

Legend and Experience

DURING the past 30 years the industrial world has spent more than $200 billion in attempts to produce useful energy from nuclear fission. Many of the seminal events and personalities in this unprecedented effort have already passed into legend.

There was the drama of the converted University of Chicago squash court where, under the leadership of the virtually canonized Enrico Fermi, a group of scientists operated the world's first atomic reactor. There was the poetry of J. Robert Oppenheimer's recollection of the ominous lines from a sacred Hindu text two and one-half years later at Alamogordo.

> If the radiance of a thousand suns were to burst at once into the sky, that would be like the splendor of the Mighty One. . . . I am become death, the destroyer of Worlds.[1]

And, of course, there was the mushroom-shaped cloud over Hiroshima. This was possibly the most powerful image of the century, perhaps even of the millennium.

A persistent, intimate association between legend and experience is the most obvious theme in the story of nuclear fission. Only slightly less obvious has been the extravagance of prophecy—prophecy of hope as well as despair. So very much has always seemed to be at stake; yet it has been difficult at any one time to disentangle image from reality and promise from performance.

Cheap electricity has always seemed to be one of the most important promises of nuclear fission. In 1945, soon after World War II ended, J. Robert Oppenheimer told a national radio audience that "in the near future" it should be possible to generate profitable electric power from "controlled nuclear chain reaction units" (reactors). He saw no real limitation in the availability of nuclear fuel; therefore, "we may look confidently to the widespread application of such sources of power to the future economy and technology of the world." That same year, the National Association of Manufacturers brought together a panel of experts to discuss the prospects for "peaceful uses of atomic energy." Dr. James B. Conant, President of Harvard University, acting as moderator, asked how long it would take to develop an atomic power plant which could compete with coal in the production of electricity. A professor of nuclear physics said that the job could be done in three to ten years, an estimate with which the President of Du Pont agreed. Other predictions ranged from 15 to 25 years.[2]

Some 30 years later, these prognostications appear to have been confirmed. Indeed, for many the promise of cheap electricity from nuclear fission became reality in 1963. In December of that year, an electric power company in New Jersey announced that it had come to terms with General Electric for the construction of a nuclear power plant which would generate cheaper electricity than a coal-burning plant. In the decade following that announcement, an impressive body of evidence accumulated in support of the hopes of those in business and government who had worked to make "peaceful" atomic energy a success. The prophecies of hope about the benefits of atomic energy to society appeared to have been fully vindicated.

In the United States the first electric power plant prototype reactor began operation in 1957 at Shippingport, Pennsylvania. It was a very different machine from the first reactor in Chicago. That first device and others built by the government as part of the wartime atomic bomb effort—the Manhattan Project—were called "piles," for they were little more than piles of graphite blocks into which uranium rods were inserted. The Chicago "pile" was not designed to produce anything; its purpose was merely to demonstrate the feasibility of starting and controlling a nuclear chain reaction. The graphite slowed down or "moderated" the neutrons produced by the splitting uranium in order to sustain and to control the process. The Manhattan Project's later "piles" also used graphite as a neutron moderator, but they were designed to produce plutonium, a new element for use in nuclear bombs.

The reactor at Shippingport had a different purpose: to turn the heat of fission into electricity. Its uranium fuel sat in a steel chamber through which ordinary water was circulated at high pressure. An intricate system of pipes, valves, and pumps allowed this circulating water to moderate the neutrons and to carry heat from the fuel chamber (the "reactor core") to an electricity-producing steam turbine.

The second American reactor to produce electricity was based on a similar design. It also used ordinary water as a neutron moderator and as a coolant. But instead of maintaining the water in the reactor core under high pressure, it permitted it to boil off into steam.

Naturally enough, these designs were dubbed "pressurized water" and "boiling water" reactors. The former was marketed by the Westinghouse Corporation, the latter by General Electric. Together, they are known as "light water reactors." The term "light water" distinguishes them from yet another design that uses a rare compound of oxygen and a form of hydrogen called deuterium to transfer heat from its core. The descriptive name of this compound is "heavy water," and devices using it are known as "heavy water reactors."

Pressurized water reactors, boiling water reactors, and heavy water reactors by no means exhaust the technical possibilities for generating electricity from nuclear fission. Since World War II dozens of other systems have been tried around the world. The first reactor to be connected to an electricity network—in England in 1953—used graphite piles as its moderator and carbon dioxide gas to remove heat from its core. During the 1950s, such "gas-graphite reactors" were the basis of British and French efforts to produce commercial nuclear energy.

The plant at Shippingport began operation in December 1957. The second American power plant prototype light water reactor ("Dresden 1," near Chicago) began operation more than one year later, in 1959; the third ("Yankee," in Rowe, Massachusetts) in 1960; and the fourth and fifth ("Indian Point 1," on the Hudson River, and "Big Rock Point" in northern Michigan) in 1962. The combined generating capacity of these five pioneer light water plants was less than 800 megawatts (MW).[3] This was a tiny fraction of total American electricity generating capacity (less than 1 percent); it was also less than 50 percent of the world's total nuclear-electric generating capacity.

Yet, in less than a decade, the closely related technologies demonstrated by the Shippingport and Dresden plants and the marketing strategies of their American manufacturers—Westinghouse and General Electric—seemed to stand as a triumph in technological innovation. By the early 1970s, affected business leaders and public officials in most industrialized countries and in many developing countries agreed that nuclear fission would be the world's principal new energy resource for the rest of the twentieth century.

There was considerable evidence to support this expectation. By the end of 1975, 157 nuclear power plants with a generating capacity of approximately 72,000 MW, or 72 gigawatts (GW), were operating in 19 countries.[4] Of these nations, all except the Netherlands and Pakistan had more commercial power reactors under construction. The People's Republic of China was reported to have already constructed one or two power reactors, and an additional nine countries had their first commercial power reactors under construction. Many other countries had announced plans to acquire reactors for the generation of electricity. An impressive record. The scientists, engineers, public officials, and business executives who worked to make nuclear power a reality might well have taken pride in bringing about one of the century's most important innovations. And as Dr. Glenn Seaborg, one of the American leaders of the effort once put it, "They did it just in the nick of time."

A worldwide conference on nuclear power was held in Paris in April 1975. Its theme was "The Maturity of Nuclear Power." A. E. Hawkins, Chairman of the United Kingdom Central Electricity Generating Board, believed this choice of theme "both commendable and appropriate." He noted that in the early days there were doubts about solving the technical problems associated with nuclear power. There had been even greater uncertainties about its economic viability, but now "all of these questions had been resolved." Nuclear power had indeed "matured."

Among all countries the one most deeply committed to nuclear power was the United States. By December 1976, only 19 years after the Shippingport reactor began operation, a total of 60 reactors with an aggregate capacity of 43 GW were completed. An additional 146 were under construction or on order. At that time the total capacity of the nation's generating network was approximately 500 GW. The enormous benefits of nuclear power were reflected in an early 1975 *Public Utilities Fortnightly* survey of all American utilities that operated nuclear power plants as part of their electrical generating systems. The 24 companies concluded that "the peaceful atom" had saved their customers more than $750 million in their 1974 bills that they would have owed had their electricity come from fossil fuels. They also reported that in the same year "power from the atom" had saved "the equivalent of more than 247 million barrels of oil."[5]

These developments were all the more striking because they were the fruit of only one reactor design—the "light water" system originally developed in the United States. By the end of 1975, this American technology completely dominated the world market for nuclear power plants. Only the Canadian heavy water system—nicknamed "CANDU"—survived as potential competition. Light water's success was achieved in spite of considerable efforts to develop and market other technologies in Europe and the United States. The most significant of these efforts was the 15-year attempt by Britain and France to make the gas-graphite design the basis of the world's nuclear electricity programs. But the last gas-graphite plant was sold by the French Atomic Energy Commission to Spain in 1967, and 10 years later no manufacturer offered such reactors for sale. Figure I–1 indicates the extent of light water dominance in the Western world.

The market dominance of American technology has been matched by the success of its two principal manufacturers, Westinghouse and General Electric. By the mid–1970s, the two American companies and their foreign subsidiaries were winning more than two-thirds of the nuclear power plant bids in non-Socialist countries.

In spite of the dual triumph of nuclear power and American reactor technology, the early prophecies of despair about the meaning of nuclear fission are not forgotten. In 1970 Alvin Weinberg, one of the early scientific and administrative leaders of the American reactor development effort, contributed, perhaps unintentionally, a chilling phrase to the legend of nuclear power:

> We, nuclear people, have made a "Faustian Contract" with society: we offer an almost unique possibility for a technologically abundant world for the oncoming billions, through our miraculous, inexhaustible energy source; but this energy source at the same

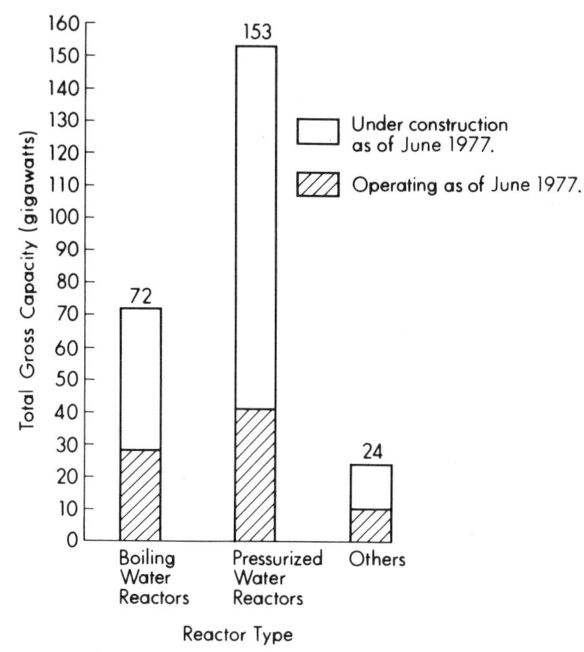

Figure I–1

time is tainted with potential side effects that if uncontrolled, could spell disaster.[6]

Weinberg's apparent warning found a receptive audience. By the early 1970s many had come to believe that nuclear fission could never be an acceptable solution to world energy problems.[7] Criticism of nuclear power began as an organized movement in the United States in the mid–1960s, almost concurrently with the initial commercial triumphs of light water technology. As those triumphs were repeated at home and abroad in the following years, criticism of nuclear power grew apace. It was not diminished by the 1973 oil embargo, which, to many, gave nuclear power the new mission of defending the interests of oil importing countries against the increasingly unpredictable oil producers.

In the spring of 1977 thousands of persons demonstrated

against the construction of a nuclear power plant at Seabrook, New Hampshire. Several hundred occupied the plant site and were arrested. Almost simultaneously, on the other side of the Atlantic at Bilbao, Spain, some 100,000 persons demonstrated against nuclear power. Although these events were not accompanied by outright violence, antinuclear demonstrations have not always been peaceful, especially in Europe. In Germany in 1976, more than 200 demonstrators were injured as they fought with police at Brokdorf. In July 1977, a demonstration against the construction of a nuclear plant at Creys-Malville, France, produced one death and several hundred injuries.

Such events are only the most visible and dramatic manifestations of a phenomenon which has become evident in all western countries with nuclear programs: a complex, often highly theatrical, confrontation between two broad sets of interests. On one side stands an array of governmental, commercial, and academic proponents committed to the rapid construction of nuclear generating plants and the industries and services they require for support. The scientists, engineers, and government and business leaders who have for 30 years shared this objective constitute an impressive concentration of public and private interests.

Opposed to this group is an equally diverse array of individuals and organizations opposed to nuclear power. In the political democracies of Western Europe, North America, and Japan these two groups are at war. In early 1978, the state of that war increasingly resembled a stalemate in which neither side seemed able decisively to impose its views about nuclear power on the community at large, due to the power of its opponent.

But, this stalemate was, in itself, a threat to something which all but an apparently tiny fraction of informed persons in all of these countries believed: that nuclear power has an important, and in some countries, crucial role to play in meeting energy needs for at least the rest of the present century. In early 1978, it seemed probable that the practical consequence of the emerging stalemate would be to limit the contribution of nuclear power to levels significantly lower than planners in both government and industry all but unanimously expected as recently as 1974.

This book is an interpretation of how the nuclear dream which emerged at the end of World War II and seemed to be on the verge of realization in the aftermath of the 1973 OPEC embargo dissolved during the ensuing four years. The central argument is that the origins of this disappointment can be found in circumstances which span the preceding 25-odd year period. Much of the first third of the book reviews events and developments of the 1950s and 1960s to answer the question of how American light water reactor technology overwhelmed the Western world market for nuclear power plants. We believe that our answer to this question contains the key to understanding the real condition of nuclear power today. To compress our argument into its most concise form: the way that the innovation process for light water reactors was managed by business and government in the U.S. and Western Europe contributed to the identification of nuclear power technology with something that many citizens in these countries dislike and distrust about their societies. Moreover, it is this dislike and distrust which is the driving force behind the nuclear safety controversy and the principal cause for the dissolution of the nuclear dream.

Part I

A TIDAL FLOW

CHAPTER 1

Euratom: A Trojan Horse for Light Water in Europe

IN AUGUST 1955, leaders of the international scientific establishment gathered in Geneva under the auspices of the United Nations to exchange views publicly on "the peaceful uses of atomic energy." A surprisingly large Soviet delegation enhanced the apparent significance of the conference, which many considered one of the most important scientific meetings of the century.

Dr. Homi Babha, Chairman of the Indian Atomic Energy Commission and Chairman of the Conference, opened the proceedings by proclaiming that nuclear power was destined to play an important role in both developed and developing countries. This prediction seemed confirmed in the opening session when the delegates learned of British plans to build some two million kilowatts of nuclear capacity by 1965. British scientists told the conference that by 1975 almost half of the United Kingdom's electricity needs would be supplied by "the peaceful atom."[1]

The British announcement had a considerable impact. Nuclear scientists and engineers from both sides of the Iron Curtain took it as a sign that the heavy fog of secrecy under which nuclear research had proceeded since the war was finally lifting. The ensuing discussions contributed to the air of euphoria. Although some papers showed that several nations, including the Soviet Union, had nuclear power research and development programs well under way, leadership was generally conceded to the United Kingdom. Everyone agreed it had the most extensive technical capabilities in nuclear power. In contrast to the United States, where the Atomic Energy Commission's development program seemed to be pursuing a number of reactor designs simultaneously, the British program concentrated upon a single concept, the gas-graphite design nicknamed "Magnox."

The progress of the British atomic energy program in the ensuing years seemed to confirm the expectations of Geneva. In the United States, congressional critics of the Atomic Energy Commission's "Cooperative Power Reactor Development Program" frequently cited British success with the gas-graphite design as proof of American tardiness. By the late 1950s many observers in both the United States and Western Europe saw the British Magnox program as the Western world's most advanced civilian nuclear power effort. They were convinced that this design offered the most certain prospect for a sizable, imminent nuclear contribution to the world's electric generating capacity.

But they were mistaken. By the late 1960s, all Western European countries except England had chosen the American light water reactor for their nuclear power programs. Their choice was justified by the argument that a major breakthrough in power reactor development had occurred a few years earlier in the United States. The reasons for the European belief in the emergent technical and economic superiority of American light water reactors is to be found in a complex set of interrelated events which took place on both sides of the Atlantic during the years preceding and following the conference at Geneva. The story begins in the United States.

"Atoms for Peace"

The Atomic Energy Act of 1946 terminated nuclear collaboration between the United States and its wartime allies, Britain and France. The bill specified that until Congress declared by joint resolution that "enforceable international safeguards" against the use of atomic energy for destructive purposes had been established, there could be no exchange of technical information with other countries for the industrial or commercial use of atomic energy. No such joint resolution was ever proposed.

The original legislation was amended in 1951 to authorize the Atomic Energy Commission to enter into arrangements with American allies for the purpose of communicating certain nuclear information. The intent of this amendment, however, was to strengthen the North Atlantic Treaty Organization (NATO) rather than to foster the commercial development of nuclear power. In fact, the overriding objective of American atomic energy policy in the postwar period was military and strategic. The government was principally concerned with retaining its monopoly of the technology for producing nuclear explosive materials.[2] This policy had crucial implications for the development of nuclear technology for nonmilitary purposes, particularly for the production of electricity. Figure 1–1 illustrates the linkages between nuclear weapons material production and electricity production. This is a complicated business, but some grasp of it is necessary in order to understand the beginning of the light water story.

Two different technologies have been developed to produce nuclear weapons. The first (labeled route 1 in Figure 1–1) requires a material known as "enriched uranium" and the second uses the element plutonium. Both of these materials must be manufactured. Uranium, as found in nature, consists of two forms or "isotopes" of different weights. The heavier,

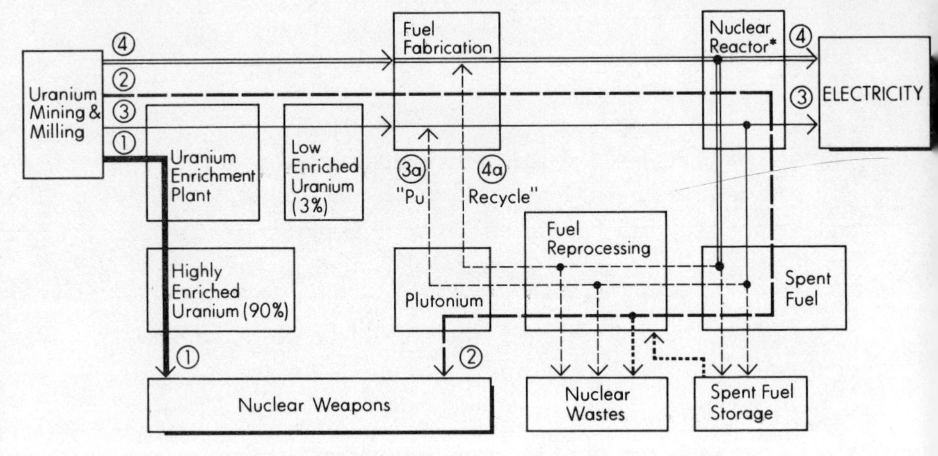

* Note that reactors fueled with "less reactive" natural uranium require a much larger size for the same electricity capacity.

Figure 1–1

whose chemical symbol is "^{238}U," constitutes more than 99 percent of natural uranium; the lighter "^{235}U" constitutes only about 0.7 percent. But it is the scarce isotope that is "fissile" and hence can be used for nuclear explosives.[3] High concentrations of this rare material are obtained through a complex process that uses the differences in the rate at which gases of varying weights filter through a porous barrier. The technology to accomplish this was developed by the Manhattan Project and it was an extremely sophisticated and expensive accomplishment. Its technical details were one of America's most important postwar atomic secrets.[4]

A second road to nuclear weapons (labeled route 2 on Figure 1–1) avoids complex uranium enrichment technology by using reactors fueled with natural uranium to produce an element not found at all in nature—plutonium. France, and later, India relied exclusively on the second route to produce their first nuclear weapons because of the American monopoly on uranium enrichment technology which it refused to share with wartime allies except England. The insistence of the

American government on retaining control of this technology obliged foreign efforts to produce electricity from atomic energy either to use the natural uranium route (route 4 in Figure 1–1) or to count on the United States to supply enriched uranium fuel. The French and the British chose the former option and developed a reactor design which used natural uranium as fuel. American scientists, meanwhile, began experimenting with designs that used enriched uranium, one category of which was the light water systems. Both enriched and natural uranium fueled reactors produce plutonium as a byproduct of operation. This, too, can be used as a fuel to generate electricity (routes 3a and 4a). Alternatively, it can be used to make nuclear weapons.[5]

The atomic energy policy embodied in the 1946 Atomic Energy Act was not an outstanding success. Notwithstanding American attempts to control all nuclear information, the Soviet Union detonated a nuclear device in 1949, and by the early 1950s the United Kingdom was well along on its own weapons program. In 1953, the new Republican administration found itself under pressure from scientists, business leaders, and diplomats to revise the nation's atomic energy policy. It had become clear that no policy of secrecy could work to prevent the spread of atomic information.

President Eisenhower's historic "Atoms for Peace" speech at the UN General Assembly in December 1953 established the policy and the conceptual framework which would control official United States thinking about atomic energy for the coming quarter of a century. All in all, "Atoms for Peace" was a quite remarkable public policy innovation. As popular as its slogan was catchy, it combined complex and subtle motivation with equally complex objectives. Doubtless it was seen as a crucial part of American efforts to manage the international military and political equilibrium. Still, underlying the rhetoric about sharing the fruits of American scientific and technical expertise, there was a hard core of pragmatic political reasoning: many of the United States' foreign policy objectives could be advanced by promoting nuclear energy.

For the Eisenhower administration one such objective was to control nuclear power development in Western Europe for greater security from the Russian military threat. Later aims promoted the development of American nuclear reactor technology as an aid to European economic and political integration.

Not the least remarkable thing about "Atoms for Peace" was its long life. Although the original slogan passed from use and the rhetoric changed, the policy's basic intent survived the next four administrations. President Carter, in 1977, became the fifth American President to reaffirm the policy formulated by Eisenhower before the General Assembly in 1953.

The United States Congress completely revised the 1946 Atomic Energy Act within several months of Eisenhower's speech. A new federal law opened the way for foreign electric utility companies and governments to build "demonstration" nuclear power plants in cooperation with American manufacturers. Shortly thereafter, the United States Atomic Energy Commission created special incentives for foreign investment in similar projects.[6]

By 1956, the Atomic Energy Commission and the Export/Import Bank had agreed on joint action to help finance demonstration reactors in countries which had entered into bilateral "Agreements for Cooperation" with the United States Government. These loans could only be used to buy equipment, materials, and technical services from the American nuclear industry. At the end of 1955, more than two dozen bilateral conventions had already been signed with countries belonging to the Organization for Economic Cooperation and Development. The standard contract provided for the delivery of enriched uranium fuel and for information pertinent to the construction and utilization of research reactors.[7]

Early Nuclear Policy in Europe

"Atoms for Peace" served multiple interests in Western Europe as well as in the United States. In the early 1950s, hopes were high that a "United States of Europe" might finally put an end to centuries of rivalry and warfare on that continent. After the French government's veto of participation in a "European Defense Community," supporters of a united Europe identified cooperation in the development of a nuclear power program as the principal remaining vehicle for realizing their objective. Because nuclear reactor technology was still in its infancy, they reasoned, it ought to be free of the well-entrenched national interests which impeded integration in other fields.[8] These rather naive views largely ignored the obvious military and industrial implications of reactor development. Nonetheless, formal discussions of a "European Atomic Energy Community" (Euratom) commenced in Brussels in July 1955. The program's proponents soon discovered that the interests and the capabilities of the prospective participants diverged sharply.

The United Kingdom, already a member of the "atomic club," announced that it could not participate in any organization with supranational authority. West Germany, which had just been allowed to start a civilian nuclear research program, was eager to capitalize on American assistance and make up for lost time. Having already started bilateral negotiations with both the British and American governments, Belgium was committed to deliver both countries substantial quantities of uranium ore from its large reserves in the Congo. Italy seemed the most likely potential client for actual nuclear power plants: its electricity was the most expensive in Western Europe. An intense British-American competition had already begun for its business.

Only France shared with Britain any significant experience

with nuclear power. In 1955, the French national nuclear research and development budget was several times larger than the total of the budgets of the other five future Common Market countries combined.[9] Because of its importance to the light water story, we will briefly review the history of early atomic energy developments in France.

The French Nuclear Power Program

In October 1945, General de Gaulle, President of the Provisional Government, had established the Commissariat à l'Energie Atomique, the French Atomic Energy Commission. He gave the new agency responsibility for the conduct of scientific and technical research on all aspects of nuclear energy including military applications. Like its counterpart in the United States, the French Atomic Energy Commission was a powerful administrative body capable of acting with considerable autonomy in domestic as well as foreign policy arenas. Rather than being attached to an existing ministry, the Commission was placed under the direct authority of the Prime Minister. Internally, scientific matters were the responsibility of a scientist known as the High Commissioner while an Administrator-General assumed all administrative responsibilities. These two officials, together with the Prime Minister (or his representative), and three other persons (in the early years, all scientists), constituted the executive body nominally responsible for the Commission's policy.[10]

In the early postwar years, scientific considerations dominated French nuclear development activities. A nucleus of prominent scientists under the leadership of the Nobel prize-winning Frédéric Joliot-Curie controlled the decision-making process and concentrated on developing the resources needed to support long-term nuclear research. But there was ambiguity and confusion about the purpose of the Commission from the start. De Gaulle's main motivation for creating the new agency was to build a French nuclear bomb.[11] But for the scientists who had been involved in nuclear research before the war, the potential civilian benefits of nuclear fission were at least equally important.

The influence of the scientists began to decline in the early 1950s as controversy developed over the Commission's program objectives and priorities. In April 1950, Joliot-Curie was dismissed as High Commissioner because of his affiliation with the Communist Party. Influential members of the government and civil service had begun to favor a stronger military orientation to the French nuclear program. They failed to secure their own candidate as Joliot-Curie's replacement, but his eventual successor, Francis Perrin, took office with a considerably diminished mandate. The Administrator-General became the agency's effective director.[12]

The new French priorities soon became clearer with the announcement of plans to build two carbon dioxide-cooled, graphite-moderated, natural uranium-fueled reactors at Marcoule. The reactors were ostensibly for scientific and industrial research, but it was an open secret that the Marcoule program was heavily influenced by military considerations. After all, the gas-graphite reactor design was one of the easiest ways to produce large amounts of plutonium for atomic weapons.

By the mid–1950s an acrimonious controversy had developed within the scientific and technical establishment between advocates and opponents of a national effort to manufacture nuclear weapons. The European Defense Community negotiations crystallized the issue of the freedom of action of the French government in nuclear weapons development and deployment. To French political leaders it seemed obvious that creation of a European Defense Community would prevent France from initiating a military nuclear program.[13]

In May 1955, the French Atomic Energy Commission and

the Ministry of Defense signed an agreement which constituted the first formal step toward nuclear weapons development. The agreement transferred substantial military funds to the Commission and scheduled construction of a third plutonium-producing reactor at Marcoule. Two months later, the government formally decided to build a nuclear submarine. Although it had not yet announced any intention to construct a nuclear bomb, it was widely recognized that a French bomb was inevitable. The government finally gave the formal go-ahead for its construction in April 1958.

The lack of explicit objectives for the French nuclear program during this entire period obscured its true direction. Many Atomic Energy Commission scientists continued to believe that the agency's chief goal was to use nuclear fission to generate cheap and reliable electrical power. The symbolic connection of the Marcoule reactors to the national electric grid in 1956 was a sign that many in the French atomic energy establishment still gave that objective high priority. But the important point about all of these developments is that although the formal decision to build an atomic bomb was not made until 1958, the major technical decisions in the preceding years purposely left the option fully open.

The Effects of the Ambiguous French Nuclear Program

The ambiguous French situation had two important effects on reactor development in Europe. On the one hand, it gave France a strong position in the Euratom negotiations. On the other, it isolated the country from the other members of the European community.[14] Prominent French proponents of European political integration—notably, Jean Monnet and

Louis Armand, President of the French National Railroad Company—argued for participation in Euratom. Guy Mollet, who became Prime Minister in January 1956, shared their views. But their opponents had persuasive reasons for shunning international cooperation.

An especially troublesome issue was the possibility that Euratom might control all fissionable material in Europe. A proposal providing for such control had, in fact, been adopted during the first meeting of the Committee for the United States of Europe arranged by Jean Monnet. Such regulation would, of course, have effectively prevented a national military program among any of the members of Euratom. The architects of the French nuclear effort were by no means prepared to abandon what they considered a key to French political independence in exchange for vague promises of cooperation from inexperienced partners. Many other influential persons in France wondered whether it was in the nation's interest to help Germany expand its nuclear capabilities. Fear of a strong German nuclear program was understandably considerable throughout France. Michel Debré, who later became General De Gaulle's first Prime Minister, contended that Germany would use the technology France would bring to Euratom and that a European Atomic Energy Community would inevitably be dominated by Germany.[15]

For Louis Armand the answer to such fears was straightforward:

> Without Euratom, we will suffer either the continued rule of bilateral agreements or the rule of an international agency, and European countries will not be anything but satellites of the large nuclear countries, the most sophisticated of which would undoubtedly be Germany. With Euratom the German potential will work for Europe and France will confirm its current advantage. . . . Our country will become a founding partner in a strong joint venture.[16]

Armand's views ultimately prevailed, but unanimity was never reached. The French government eventually signed the Euratom Agreement, suspicious of its future partners and with deep reservations about the objectives of the new organization.

The 1956 Suez crisis reinforced the widespread conviction that rapid development of nuclear power was essential for Europe. In November 1956, the governments of the six future Common Market countries appointed three prominent "pro-Europeans," including Louis Armand, to study the future energy requirements of the European Community. Newspapers quickly dubbed the trio "the Three Wise Men." The other two were Franz Etsel, the German Vice President of the European Coal and Steel Community, and Francesco Giordani, former President of the Italian Atomic Energy Commission. With considerable fanfare the Three Wise Men visited the United States, the United Kingdom, Canada, France, and other European countries to assess the prospects for nuclear power. Their subsequent report, *A Target for Euratom*, stated that the European Community urgently needed new energy sources. The report recommended the installation of 15,000 MW of nuclear generating capacity—equivalent to about 25 percent of the Common Market's total electrical supply system—by 1967. It also advocated close cooperation with the United States.[17] In doing so, the Wise Men challenged the prevailing consensus that the gas-graphite technology developed by France and England was the most advanced or "proven" reactor design. According to them, American light water reactor technology was equally advanced. In general, *A Target for Euratom* argued that the nuclear unification of Western Europe would quickly open the Continent to the superior technology of the American nuclear industry.

This enthusiasm for American nuclear experience was not shared by all the members of the European Economic Community. An alternative conception, supported by many in France, advocated collaboration by the six European countries to develop an indigenous technological base in nuclear power.[18]

Nevertheless *A Target for Euratom* was formally delivered to the member governments on May 4, 1957, and its conclusions were influential in persuading some of them that they needed Euratom. The feeling was heightened when, late in 1957, the British Ministry of Industry announced it was tripling the already ambitious 1955 British nuclear program. The parliaments of all six member countries of the European Economic Community signed the Euratom Treaty during the spring of 1957.

By 1958 most European countries, including France, had signed bilateral arrangements with the United States for the delivery of limited amounts of fissile material for nuclear power research.[19]

In Italy, cooperation with the U.S. began as the result of a World Bank study which concluded that Italy and Japan were the countries most likely to develop competitively priced nuclear power in the near future. The Bank suggested that the Italian government undertake a study of the possibility of building a 150 MW nuclear power plant. In the summer of 1957, after a number of meetings with the Bank's representative, the Italian government finally agreed on terms for the project's development. Bids were sent to 15 American, British, French, and Canadian companies, and nine replied by the closing date of April 30, 1958. An international panel judged the bids and awarded the contract to an American proposal for a light water reactor whose electricity would be only 10 percent more expensive than that from conventional sources.[20]

By 1958, then, bilateral agreements had already allowed American technology to penetrate the major European reactor markets. Influential Europeans strongly supported these arrangements. Among them, Louis Armand remained the most enthusiastic about American technological prowess. The origins of America's superiority were, to him, self-evident and inherently linked to the organization of American society itself.[21] In Armand's view, the European countries could never equal American technical capabilities.

> One can see no country, even France, with its three reactors of the same type under construction . . . that could be capable alone of counterbalancing the U.S. which already has 30 different types of reactors manufactured.[22]

Commenting on a 1956 speech by Armand, Bertrand Goldschmidt, a pioneer of the French atomic energy program, observed:

> Armand's speech was a panegyric to the maxim that strength increases through union; he insisted on the lag of European countries, vis-a-vis the U.S. and Soviet giants, while minimizing the capacity for national programs by recalling among other things the failure of the British Comet and the success of the U.S. Boeing Aircraft.[23]

It was, therefore, with a strongly favorable bias, to say the least, that Armand and the other Wise Men had gone to North America. They were warmly received. The U.S. Atomic Energy Commission detailed four members of its staff to accompany the Europeans throughout their trip. Apparently, these American officials served as rather more than simple technical advisors. Among other things, they reportedly drafted the final version of *A Target for Euratom*.[24]

> The influence of these experts was extensive. They supplied technical data for the energy projections and assured the Three Wise Men of full U.S. support in meeting nuclear construction goals. They made rosy forecasts of the economic performance of U.S. light water reactors which were still under construction. Euratom had no technical staff nor independent data with which to challenge these American estimates.[25]

No wonder, then, that the Wise Men's report was so favorable to American technology.

Industrial development of nuclear power in the United

States had hardly begun in 1958. Shippingport had opened its doors only a little more than one year before, and prospects for economically competitive nuclear power were still highly uncertain. Many persons in the American nuclear community believed that nuclear power would become competitive in Europe far earlier than in the United States. The proposals of the Three Wise Men meant an exceptional opportunity for American manufacturers: Europe could become a proving ground for American light water technology.[26]

Shortly after its creation, the Euratom High Commission established a working group with the American government to develop a reactor prototype construction program. In November 1958, an agreement was signed calling for the equivalent of one million kilowatts of nuclear power to be built (under American patents) between 1960 and 1968 in Euratom communities. The cooperative arrangement provided for low interest Export/Import Bank loans, lease of enriched uranium fuel from the U.S. Atomic Energy Commission together with U.S. government guarantees of the supply and performance of the fuel, and for a jointly funded research and development program. The direct and indirect American government contribution to Euratom's research and power plant construction program totaled about 50 percent of its estimated budget of $350 million.[27]

The Euratom Treaty completed the foundation for what would become an imposing edifice in little more than 10 years: a European nuclear energy supply system based wholly on American technology. To fully understand the significance of these events, we must review contemporary development in the United States.

The U.S. Reactor Development Program

The U.S. reactor development program in the 1950s bore slight resemblance to the prevailing European conception of it.[28] A commercially useful nuclear reactor was not a high priority for the Atomic Energy Commission in the first several years of its existence, in spite of periodic prodding by J. Robert Oppenheimer and the Commission's very prestigious General Advisory Committee. The agency's administrative and fiscal resources were far too absorbed by the requirements of the weapons program. By 1947 the production of plutonium for atomic bombs had dropped to a fraction of its wartime rate, and further declines anticipated. Sustained operation of the three plutonium-producing plants at Hanford, Washington had caused expansion of the moderating material, and the Army had been forced to shut down the oldest reactor in 1946. The two remaining plants were operating at greatly reduced power to conserve their lives.[29]

Yet, by the summer of 1949, the Atomic Energy Commission managed to start four loosely related projects: a submarine propulsion program jointly administered with the Navy; a "materials testing reactor"; an experimental advanced reactor at the Argonne National Laboratory; and another experimental project at the General Electric Company's Knolls Atomic Power Laboratory. The detonation of the Soviet Union's first nuclear bomb in August 1949 caused all four projects to be increasingly oriented to military needs. By the spring of 1950, almost all government-supported work on reactor development was directly geared to military projects.[30]

Only General Electric remained heavily committed to a strictly civilian project. Serious technical difficulties had, however, arisen with its "intermediate breeder reactor" design.

In "breeder reactors," at the same time that the initial fissionable fuel charge is being used up, even greater amounts of a new fissionable material—plutonium—is being created from the ^{238}U which is also in the reactor core. Plutonium is also "bred" in relatively modest amounts in light water reactors. But in breeder reactors this takes place at a far faster rate. The attraction of using such machines as a power source is that the amount of energy available from natural uranium is in principle more than 100 times greater than that available from light water designs. The construction of such breeder reactors has been an almost unanimous ambition of civilian nuclear scientists since the war. Only breeder reactors would fulfill the scientific dream of nuclear power as an unlimited energy source. However, the early General Electric design encountered serious problems. In order to keep alive any hope of breeding plutonium, company scientists had been forced to redesign their reactor for higher neutron energies that severely restricted its potential as a power producer. In the early years of reactor development, the goal of a single plant that could generate electrical energy while simultaneously replenishing the supply of fissile fuel looked increasingly difficult to achieve.[31]

The last two years of the Truman Administration saw a steady growth of activity on military reactors, first for submarine and subsequently for aircraft propulsion. The administration gave highest priority, though, to increasing the supply of fissionable materials for explosives. In early 1952, the National Laboratories' chief mandate was to improve the design of reactors whose sole purpose was to produce plutonium for bombs. A plutonium-producing reactor which could also produce cheap electricity was a secondary objective.

Meanwhile, the laboratories and the AEC's industrial contractors had been making progress in reactor construction. The experimental breeder at Argonne operated at design power for an extended period in 1952; the materials testing reactor went critical on March 31, 1952; and by November of that year, the first prototype for a submarine propulsion reactor was essentially complete and a second was well under way.[32]

But still there was no Atomic Energy Commission project whose principal purpose was commercial nuclear power. In the face of growing technical problems and persistent pressure from the Commission, General Electric had canceled its breeder project in the spring of 1950 in favor of work on a submarine reactor. In April 1951, the Commission's Director of Reactor Development observed that ". . . the cost of construction of a nuclear power plant is still essentially unknown. We have never designed, much less built and operated, a reactor intended to deliver significant amounts of power economically."[33]

In January 1951 several industrial groups expressed to the Commission willingness to bring their technical resources into the national atomic energy program. Very little came of these offers beyond some joint studies of the practicality of private industry's building additional reactors for making weapons material.

The first Eisenhower budget eliminated an AEC request for funds to construct a pilot power plant, but at the urging of the Joint Committee on Atomic Energy, the House Appropriations Committee restored the money. Shortly thereafter, the Atomic Energy Commission announced that Westinghouse and Duquesne Light Company would build and partially finance a power plant at Shippingport, Pennsylvania based on the pressured water reactor technology Westinghouse had developed for the Navy. Essentially, Shippingport would be a larger version of a reactor designed for an unbuilt nuclear aircraft carrier. The Joint Committee later claimed full credit for what has since generally been conceded to have been the first major step toward commercial nuclear power in the United States.[34]

In 1954, the Joint Committee completely rewrote federal atomic energy legislation to allow private ownership of reactors under Atomic Energy Commission license. Under the leadership of Lewis Strauss, however, the Commission chose to interprete the new legislation as a mandate to rely principally on private industry to develop civilian reactor technology. This policy was a source of chronic distress to the Democratic majority on the Joint Committee who constantly tried to prod Strauss toward more aggressive action.

In January 1955, the Atomic Energy Commission announced a new "Power Reactor Demonstration Program" and offered to review competitive proposals for government-assisted, privately financed reactor projects. The AEC would, for acceptable proposals, waive all charges for the use of fissionable materials, undertake certain basic research in its National Laboratories at government expense, and enter into fixed-sum research and development contracts to procure technical and economic data for the applicants. The fixed-sum commitment placed a ceiling on AEC participation, so the applicants would bear the projects' economic risks. The Democrats on the Joint Committee did not regard these terms as generous.

Nor was industry especially enthusiastic. The Atomic Energy Commission received only four proposals. Moreover, public-power advocates raised a great hue and cry that municipal utilities and rural electric cooperatives could not share in the bounties of nuclear power because they lacked the substantial risk capital required by the government program's provisions. The Commission finally responded to the Joint Committee's displeasure by issuing a second-round invitation in September 1955. This time the Commission would consider requests for financing reactors in whole, or in substantial part, but it would retain title to that portion it financed. Seven proposals were submitted in the second round.[35]

The Joint Committee was not mollified. By early 1956 Committee Democrats were enraged by what they considered the glacial pace of Lewis Strauss' reactor development program. The United Kingdom was widely cited as having already far outstripped the United States because of his misguided policies. In January 1956, a panel of businessmen and academics appointed by Senator Clinton Anderson published a report which among other things recommended that the AEC construct a full-scale prototype of every major reactor type.[36] Commissioner Thomas Murray, a Democrat holdover from the pre-Strauss Commission, publicly called for a $1 billion government-financed reactor construction program.

Strauss remained adamant. In his view all that was needed was the removal of certain roadblocks in the path of private nuclear development—the possibility, for instance, that utilities working cooperatively might be considered a violation of federal government antitrust legislation. Another roadblock he saw was the need to protect industry against the financial liability a serious nuclear accident would entail. Congress soon passed legislation to eliminate these difficulties. Meanwhile, the Democrats' continuing dissatisfaction materialized as a bill sponsored by Senator Albert Gore and Representative Chet Holifield. The Gore-Holifield Bill would have directed the Atomic Energy Commission to construct six full-scale prototype plants wholly at government expense. After a long and heated debate, the Joint Committee reported the bill out in a substantially weakened version. An even more watered-down version passed the Senate but was killed by the House.

Although the policy battle saw periodic skirmishes throughout President Eisenhower's second term, the defeat of "Gore-Holifield" was the high-water mark of the Democrats' assault on Strauss' policy. The Democrats on the Joint Committee never effectively challenged its central premise: that the government should restrict its activities to exploring advanced reactor concepts on a pilot scale by building small experimental plants. Under Strauss, the Commission never accepted full responsibility for building large demonstration reactors. In 1957, it issued a third round of invitations, but once again the terms precluded direct government financial assistance with construction costs.

Strauss was excoriated by his Democratic colleague on the Commission, Thomas Murray:

> Our first consideration in weighing the issues of industrial atomic power should be the national interest . . . [it] demands the development of atomic electric power on an urgent basis. . . . Private industry is not prepared today to provide more than a portion which the national interest requires. . . . Industry's time schedule is set primarily by this country's needs for electric energy. But the time schedule required by the national interest is much shorter. It is set by the crisis in nuclear weapons and the world need for atomic power.
>
> . . . [The reactor development program] does not distinguish between foreign needs and domestic needs—between short-term goals and long-term goals. It leaves responsibility for designing and constructing large reactors to private industry. However, if we are to get the answers to atomic power quickly, the government cannot relinquish the responsibility for constructing large-scale reactors during the years immediately ahead.[37]

But Murray's view of the government's proper role in the nuclear program was not shared by important elements of the American business community. Many utility executives, for example, believed that if the government fully subsidized the construction of prototype reactors it would be in the electric power business, and investor-owned companies would be squeezed out by "massive atomic TVAs."

Because of this persistent, paralyzing, political controversy, commercial nuclear power was in its infancy in the United States when the member countries of the European Economic Community signed the Euratom Treaty in the spring of 1957. Prospects for economically competitive power from the light water design were more theoretical than real. The cooperative arrangements the "Atoms for Peace" program made possible, therefore, provided an enormous opportunity for Americans to undertake projects abroad that were precluded by political forces and economic constraints at home. The direct and indirect financial assistance provided by the government to

European nuclear programs was additional incentive for American manufacturers to capitalize on the opportunities that "Atoms for Peace" handed them.

In effect, American nuclear policy during the 1950s promoted on one side of the Atlantic what it inhibited on the other. In Europe, American commercial interests benefited from direct government financial support for the development of light water reactors at a time when the Atomic Energy Commission's domestic reactor development policy inhibited such support at home.

The Results

In April 1959, under the terms of the United States/Euratom Bilateral Agreement, Euratom invited proposals to build prototype reactors. The invitation provided for:

1. Financial guarantees of the cost and integrity of the nuclear fuel for a 10-year operating period;
2. Long-term assurance of an adequate supply of fuel at prices comparable to those offered to American industry. The United States also agreed to furnish Euratom with up to 30,000 kilograms of ^{235}U on credit at 4 percent interest;
3. A guaranteed market for the plutonium recovered from the spent fuel for 10 years;
4. Long-term capital loans to cover part of the cost of plant construction; and
5. American assurances that the reprocessing of spent fuel would be available to participants on terms comparable to those offered operators in the United States.[38]

By the deadline of the invitation, however, only a single proposal—from Italy—was in hand. In September 1961, the U.S. Atomic Energy Commission and Euratom tried a second time. The new invitation led to two additional projects. The first was a 380 MW plant to be built at Chooz, in France, near the Belgian border. The second was a 237 MW plant to be built at Gundremmingen, Bavaria for a consortium of West German utilities (KRB). An American firm was to be the designer and supplier of the power plant for both projects.

Arrangements for the first of these three "Euratom reactors" had begun several years before the first invitation. The decision to build a nuclear power plant at Garigliano, in Italy, had actually been made in September 1958. It is likely that the Italian determination to build a light water reactor rather than a British gas-graphite system was influenced by anticipated participation in the Euratom program with its attractive government assistance provisions. But whatever the reason for that choice, its psychological impact throughout Western Europe was great. It helped to overcome the widespread conviction about the technical superiority of the British design. For the first time in Europe, light water reactors achieved technical respectability and even, perhaps, equivalent prominence with the gas-graphite design.[39]

The plant at Chooz was a joint French-Belgian venture. Here, too, the choice of light water was politically and psychologically significant. The principal French sponsor of the Chooz plant was Electricité de France. This agency saw light water as a way to develop "in-house" capabilities in nuclear power, thereby allowing a challenge to the monopoly of the Commissariat à l'Energie Atomique.

The KRB utilities' choice of a light water design was also a major turning point in German reactor commercialization. In 1961 Germany still lacked a nuclear power plant of the size being built by the Italians, the Belgians, and the French. The bidding for this first German demonstration reactor had developed into a contest between a German/American proposal for a 200 MW light water system and a British plan for a 300–400 MW gas-graphite plant. Only the light water reactor was eligible for assistance from the Euratom cooperative program.

As early as 1960, the largest German electric utility (RWE)

had become convinced that light water plants would be more economical in the long term than British gas-graphite systems. In fact, this belief was evidently common throughout German commercial circles about that time; analysts chiefly cited the lower initial investment costs of light water reactors relative to gas-graphite systems for plants of equal capacity. But the influence of the Euratom cooperative program cannot be overlooked. As a close student of these developments, Professor Henry Nau, put it: "The program altered expectations in favor of American reactors from 1958 onward."[40] More concretely, by participating in the American light water program, Europeans were assured of a hand in American research results and enriched uranium supplies. This second advantage removed a potentially decisive obstacle to light water technology in Europe.

When the KRB utilities made their choice, no light water plant of the size they contemplated was operating in the United States. But, their decision influenced the principal German reactor manufacturer, Siemens, to reverse its own development priorities in favor of the American technology. Siemens' turnabout against gas-graphite came at almost exactly the same time that France and Germany began discussions on the joint construction of a 500 MW gas-graphite plant in Fessenheim, France. Siemens represented Germany in these talks, and its loss of interest in gas-graphite systems hardly improved prospects for cooperation with France. When Germany ordered two additional light water plants in 1964, these prospects—never more than modest—were altogether dashed.[41]

Under de Gaulle, the French nuclear program had become a program to produce nuclear weapons.[42] The openness of his intentions only widened the already considerable divergence of purpose between France and the rest of the Common Market with respect to the future of civilian nuclear power. With military requirements as the cornerstone of its program, the power and financial capabilities of the French Atomic Energy

Commission were greatly expanded. In 1961, the Commission controlled more than 2 percent of the national budget. By 1962, the number of Commission employees had increased by a factor of more than 9 over those employed 10 years earlier.

For many of the scientists and engineers who, 10 years earlier, had begun the French nuclear power program, initial hopes were undimmed. They still considered the experimental development of gas-graphite reactors and their eventual commercialization throughout Western Europe as the basic purpose of the French nuclear program.Their views were influential in the ultimate, although reluctant, acceptance of the Euratom Treaty by the French Atomic Energy Commission.

From the Commission's point of view, only France among the six Common Market countries had serious nuclear experience.[43] Executives in the Commission generally expected that their representatives to Euratom would rapidly demonstrate the competitiveness, if not the outright superiority, of French gas-graphite technology and ultimately influence all of Euratom's choices. France's partners in Euratom, however, had strikingly different views of the common nuclear enterprise. Not only had most of them already based their national efforts on "Agreements for Cooperation" with the United States, but fear of French hegemony over the Euratom Commission was universal, and especially acute, in Germany.[44]

During the early 1960s, French participants did seriously attempt to modify the orientation of the Euratom program. They made substantial efforts to reach an agreement with Germany based on gas-graphite technology. In retrospect, it was probably the KRB choice of a light water reactor system for Gundremmingen and Siemens' subsequent loss of interest in gas-graphite technology that doomed the French strategy.

These events very quickly disenchanted the French Atomic Energy Commission. Its officials were especially distressed with the current Chairman of Euratom, M. Hirsh, whom they saw as an indifferent protector of vital French interests. In 1961 France alone vetoed renewal of Hirsh's presidency. General de Gaulle gave the new president, Pierre Chatenet, a personal mandate to reinforce French influence on Euratom policies. Chatenet, however, quickly realized that attempts to change the situation would be fruitless; by 1962 a reorientation of Euratom nuclear development programs toward gas-graphite technology had become a practical impossibility.[45]

Thus, the French attitude toward Euratom became progressively more negative and obstructive. From the French Atomic Energy Commission's point of view, France's partners in Euratom had effectively played a central role in promoting American technology throughout Europe in preference to established, indigenous gas-graphite designs. It is not difficult to sympathize with the bitterness felt by the Commission's scientists and engineers. The joint Euratom program may not have been the sole cause of the triumph of American technology in Germany and Italy. But, it was certainly easy for that impression to arise. It is equally certain that the Euratom program materially influenced, even if it did not wholly determine, the fate of British and French gas-graphite programs.

The combination of a systematic antagonism by the French toward Euratom's reactor program and a declining sense of urgent need for a massive construction effort eventually led to the collapse of the program's original grandiose plans. Instead of the 15,000 MW of nuclear generating capacity initially called for, only three plants with an aggregate capacity of 750 MW were actually completed. Yet these three plants are arguably the most commercially important nuclear plants ever built outside the United States. As a direct result of these projects, American manufacturers developed strong working relations with utilities in Germany, Italy, and Belgium; and outside France, the most important potential reactor manufacturers switched development priorities from gas-graphite to light water systems.

In the early 1960s, the fossil fuel picture in Europe was considerably brighter than it had been at the end of the prior decade. A coal supply shortage which had started in 1957 had eased. This fact, combined with a gradual decline in the price of petroleum on the world market, began drastically to change the conditions which had contributed to the previous decade's optimism about the economic prospects for nuclear power. American reactor manufacturers could plausibly conclude that the commercial development of nuclear power in Europe was further in the future than it had once appeared. The time had come to capitalize on the European experience and return to the principal market, in the United States. Here a comparatively slower decline in the cost of fossil fuels made nuclear power more attractive.

Meanwhile, in France and the United Kingdom, the only European countries still uncommitted to light water reactors, gas-graphite systems were confronted with steadily declining fossil fuel prices. When news came of an apparent economic breakthrough for nuclear power in the United States light water reactors, European reactor technology would breathe its last.

CHAPTER 2

A "Great Bandwagon Market" for Light Water in the United States

IN DECEMBER 1963, the Jersey Central Power & Light Company announced its purchase of a 515 MW light water reactor from General Electric. To explain its decision, Jersey Central offered a reason no one had ever heard before: the Oyster Creek plant would produce electricity more cheaply than any other generating system. Also unique was the non-participation of the U.S. Atomic Energy Commission. For the first time a nuclear power plant would be built without any direct subsidy. The manufacturer of the proposed nuclear power plant had offered to build the complete generating facility for a price which, between the beginning and the end of the multi-year project, could change only according to certain indices to correct for monetary inflation.

Jersey Central's decision was widely regarded as a milestone in the development of power reactor technology. It was accepted as proof that the day when nuclear power plants would be sold through the United States in direct competition with conventional plants was very close at hand, if indeed it had not already arrived.

Several weeks after its initial announcement, Jersey Central released an economic analysis of the Oyster Creek project. According to the principal trade journal of the American nuclear industry, this analysis "confirmed in the strongest possible way" that earlier economic evaluations of nuclear power had become "obsolete." Jersey Central's report, said the *Forum Memo*, established that the costs of building and operating large light water power plants were "now at levels which would have seemed incredibly low a year ago."[1]

The electric utility believed that within five years of start-up, its nuclear plant would be more economical than any conventional power source. To make a coal-fired station at the Oyster Creek site competitive with the nuclear plant, for example, the cost of coal delivered to the plant would, according to the Jersey Central economic analysis, have to have been less than 20¢ per million British thermal units (20¢/mbtu).[2] At the time, the average cost of coal consumed by American utilities ranged from a high of more than 35¢/mbtu in New England to less than 20¢/mbtu for plants close to the Appalachian coal fields. Jersey Central's own average coal costs were about 29¢/mbtu.

The assumption that the proposed nuclear power plant could be operated steadily at a power level about 20 percent higher than its guaranteed rating was important to the economic justification for the purchase. Based on a "stretched" capacity of 640 MW, the utility's analysis projected an initial construction cost of approximately $100/kw, a figure which the *Forum Memo* said would have been "almost unimaginable a year ago."[3] (The initial construction cost of a reactor, or its "capital cost," is usually expressed in $/kw. This figure is obtained by dividing the total cost of the plant by its size.)

Another crucial element in the utility's economic analysis

was the expectation of an extraordinarily high capacity factor. Jersey Central postulated that the nuclear plant could be operated at an average capacity factor of 88 percent over the first half of the plant's projected 30-year life. The capacity factor is the ratio of the amount of electricity per year the plant produces to the amount it could theoretically produce if it operated nonstop at full capacity. This ratio can never be 100 percent because of inevitable maintenance, fuel reloading, and repairs.

In February 1964, spokesmen for General Electric maintained that the low cost of the Oyster Creek plant was by no means unique.[4] General Electric was also furnishing another utility in upper New York State, Niagara Mohawk, with the major components for its "Nine-Mile Point" nuclear station at prices "in line" with those for Oyster Creek. In addition, the company intended to publish a price list for nuclear plants of many sizes. The Westinghouse Electric Company was quick to match General Electric's price quotations with its own for a slightly different type of light water system, the "pressurized water reactor." A race had clearly begun. So, too, had a rather curious "debate" about the meaning of these events.

Background to Oyster Creek: Debate without Disagreement

In November 1962, the U.S. Atomic Energy Commission sent to the White House a study titled: *Civilian Nuclear Power—A Report to the President—1962*. It quickly became a benchmark reference document for the American nuclear power industry. In it, the AEC claimed that significant progress had been made in nine years since Shippingport was authorized by Congress. It estimated that the costs of electricity generated by light water reactors had been reduced, from about 50 mills/kwh[5] at Shippingport in 1958 to less than 10 mills/kwh in contemporary plants. This cost was expected to plummet to an estimated 5.6 mills for a large plant soon to be built by the Pacific Gas and Electric Company at Bodega Bay, California. The Commission was optimistic about the future economic prospects for light water reactors, concluding that nuclear power was "on the threshold of economic competitiveness" and could "soon be made competitive in areas consuming a significant faction of the nation's electrical energy."[6] Relatively modest assistance by the government would ensure that threshold was crossed in good time.

So it was that the Jersey Central announcement, coming less than two years after the release of the AEC study, seemed to confirm official optimism about the progress of reactor commercialization. The Commission had estimated in 1962 that light water reactors entering operation in 1980 would produce electricity for only 3.8 mills/kwh during the first five years of their operation. Yet, the Oyster Creek plant, scheduled for first operation in 1968, would meet that goal years earlier.

Few informed persons in either industry or government publicly questioned this rosy picture. Philip Sporn, President of the American Electric Power Company, was virtually alone in challenging these economic analyses of nuclear power. At the request of the Congressional Joint Committee on Atomic Energy, Sporn reviewed the AEC's 1962 Report. In his view "excitement and preoccupation with nuclear fission" had produced a "disposition to sweep certain difficult or unpleasant facts connected with nuclear technology under the rug."[7]

Nevertheless, Sporn was far less critical of the estimates of nuclear electricity's cost than he was about the comparative economics of nuclear and fossil generating plants. His principal contention was that the nuclear industry and the government had been insufficiently attentive to the rate of progress in conventional generating technology. Sporn argued that fossil fuel technology was still in a state of dynamic development. Between 1900 and 1960, there had been an eightfold increase in the amount of electrical energy that could be extracted from conventional fuels, and in his view, this spectacular progress was still not at its end. He also contended that the AEC had made an "improper appraisal" of the present and future costs of conventional fuels. He pointed out that in many parts of the world the cost of both coal and oil were far cheaper in 1962 than they had been in 1952. Moreover, there was a "veritable glut" of energy in many of the technologically and economically advanced countries.[8]

The AEC criticized Sporn for his pessimism, and he received surprisingly little support from those in competitive fuel industries who might have been expected to welcome his skepticism. The coal industry focused on the secondary issue of whether it was legitimate to attempt a comparative economic analysis of the two technologies when one of them benefited from various government subsidies. We found no indication that anyone raised the more fundamental point of the uncertainty involved in making cost estimates for nuclear plants for which there was little prior construction experience and of nonexistent industrial processes necessary to use uranium in light water reactors.

Sporn himself acknowledged that his sole source of information on nuclear economics had been presentations made to his own company by reactor manufacturers. He also explicitly noted his concurrence with the two key assumptions of the reactor manufacturers' own analyses. The first of these was that it was possible to predict costs of nuclear power plants which were two to three times larger than those already operating. Second, the companies believed that "learning effects" would help reduce costs in the early years of nuclear plant construction. Sporn accepted both of these assumptions with only minor reservations.[9]

In its reaction to Sporn's analysis, General Electric objected: "We feel that Mr. Sporn's choice of a reference reactor system may have been unduly conservative. For example, the capital cost for a 460 MW plant . . . is $30/kw higher than we are presently quoting for a complete plant (of comparable size.)"[10] This difference of $30/kw between Sporn and General Electric represented less than a 15 percent difference in their estimate of the total capital cost of nuclear power plants. This narrow difference would appear to be well within the inevitable band of uncertainty for any first-of-a-kind technology. A skeptic might have supposed that reactor construction and operating experience in 1963 was hardly sufficient to allow cost predictions for new, larger plants at this level of precision, but no one seemed prepared to acknowledge this fact. Rather, the AEC agreed with General Electric that ". . . Mr. Sporn could have been more optimistic in his economic appraisal of the nuclear alternative."[11] All of the participants in the first public debate on the economic status of nuclear power agreed that there were highly predictable economic benefits to be derived from scale economies and learning by experience. These expectations were, of course, consistent with the experience of other high technology industries at that time.

The disagreements about the capital cost of reactors among the manufacturers, Sporn, and the government were on the order of 10 to 15 percent. There were few skeptics to challenge the basis for this relative certainty. The distinction between empirically supported fact and expectation—often quite obviously self-interested expectation—was blurred from the beginning in the discussion of nuclear power economics.

Nonetheless, by early 1963, there was a consensus on the estimated costs of electricity from light water reactors. The government and the reactor manufacturers were at one extreme, predicting about 5.8 mills/kwh; and Sporn was at the other, predicting about 7.6 mills/kwh—a relatively small difference of about 25 percent. Contemporary costs of electricity from fossil fuels ranged between 4.3 mills/kwh and 6.4 mills/kwh. Hence, depending upon the assumptions one chose to work with, nuclear power was either "quite competitive," "almost competitive," or "not yet competitive," in 1963.

The AEC's 1962 Report to the President had argued that dramatic developments in the reactor commercialization process were close at hand. Nuclear power would soon be able to compete with conventional generating technologies. Discussion of the AEC's conclusions revealed a range of differences narrow enough to justify serious consideration by electric utilities of the purchase of nuclear power plants.

Aftermath of Oyster Creek: A "Great Bandwagon Market"

In the year following the Atomic Energy Commission's 1962 Report, intense competition developed between General Electric and Westinghouse over several potential nuclear power plant sales. During those crucial months the two major vendors evidently decided to take drastic action to gain an assured market share in what would soon be a lucrative business. The result was a series of "turnkey" offers to build a complete nuclear generating facility at a contractually secured, "firm" price. The manufacturers committed themselves to deliver complete nuclear power generating stations at prices subject to change only according to a formula designed to reflect monetary inflation. They would assume responsibility for the cost of materials and equipment manufactured by other companies as well as the responsibility for managing the entire construction project. All the electric utility had to do was "open the door" of its complete plant at a specified date in the future and start the generating equipment—hence the name "turnkey."

The Jersey Central Power & Light Company was the first

utility to accept such an offer for its Oyster Creek plant. The cost of electricity General Electric guaranteed to Jersey Central was even less than the most optimistic estimates made less than a year previously. Within a few months the advertised economic status of light water reactors had declined from 5.8–7.6 mills/kwh to 4.3 mills/kwh without any new construction experience. In keeping with standards of skepticism established during the debate on the 1962 AEC Report, the fact that the latter figure was expectation and not accomplishment was not the object of widespread attention.

The first "turnkey" contract was quickly followed by eight others, and these sales were accepted as proof of the reality of commercial nuclear power from light water reactors.

Yet another new phase of reactor commercialization began in 1965 when American utilities placed their first orders for nuclear plants without firm price guarantees by the manufacturers. In 1966–67 a "Great Bandwagon Market"[12] for nuclear plants developed as U.S. utilities placed firm orders for 49 plants, totaling 39,732 MW of capacity.[13] Two other manufacturers—Babcock & Wilcox and Combustion Engineering—had already joined General Electric and Westinghouse in the light water reactor business. Intense competition very quickly developed among the four reactor manufacturers. This competition was waged in terms of prices of equipment and with ancillary guarantees on fuel delivery and other factors affecting future plant operating costs. As a practical matter, in most cases, a utility considering the purchase of a reactor was able to solicit secret bids from each of the four manufacturers and then to negotiate among the lowest bidders for the most attractive supplemental guarantees. This remarkable buyer's market was characterized by continuous downward revision of the estimated cost of electricity from nuclear plants. By the end of 1967, U.S. utilities had ordered 75 nuclear power plants totaling more than 45,000 MW of generating capacity. More than 80 percent of these orders were placed in 1966–1967.[14]

Proponents of nuclear power were ecstatic. Many spoke of a "revolution" having been accomplished. "Nuclear reactors now appear to be the cheapest of all sources of energy," Alvin Weinberg, the Director of the Oak Ridge National Laboratory, told the National Academy of Sciences. He saw the "nuclear energy revolution" to be "based upon the permanent and ubiquitous availability of cheap power." Among informed experts, only Sporn questioned this assessment. He saw "a bandwagon effect, with many utilities rushing ahead to order nuclear power plants, often on the basis of only nebulous analysis and frequently because of a desire to get started in the nuclear business."[15]

Nonetheless as the "Great Bandwagon Market" drew to a close in early 1968, there was euphoria in the United States nuclear industry. Both the nuclear industry and the AEC were making forecasts of installed nuclear capacity in 1975 and 1980, which would have seemed surprising to say the least to even the most bullish proponent of nuclear power ten, or even five, years earlier.

The AEC Practices Benign Neglect of Light Water Reactor Development

The U.S. Atomic Energy Commission's reactor development program underwent an historic transformation almost simultaneously with the beginning of the Great Bandwagon Market for light water reactors, but the changes in its policy and internal organization were only indirectly related to these commercial developments. Instead, they were the culmination of the decade-long squabble between it and the Joint Committee on Atomic Energy about the proper role of government in reactor development. The changes were more related to the issues of the 1950s than to the rapidly unfolding events of the 1960s and their implications for the early 1970s.[16]

In November 1964, Milton Shaw was named Director of Reactor Development for the Atomic Energy Commission. He came to the Commission from Admiral Hyman Rickover's naval propulsion reactor program. His mandate was to turn the Commission's reactor development activities into the type of aggressive, government-controlled program which the Joint Committee had been demanding since the Eisenhower years. In redefining the government's reactor development program, however, Shaw did not give top priority to the rapidly changing commercial outlook for light water systems. For both Shaw and the Commission a more important problem of reactor development was that light water technology seemed wasteful of uranium resources. The new AEC policy was based on an accelerated effort to develop more "advanced" systems. These would simultaneously use uranium more efficiently and have a higher thermodynamic efficiency by operating at higher temperatures.

Five different reactor design concepts, each associated with a particular manufacturer, had seemed to offer roughly equal promise for "second generation" nuclear power plants. Prior to Shaw's arrival, the Commission's technical staff had taken the laissez-faire position that electric utilities would have to make their own selections from this menu. In February 1964, they had solicited proposals for cooperative prototype projects without indicating the technical superiority of any one concept over the others. Several proposals were soon in hand. Throughout the spring and summer of 1964 the choice among them occupied the Commission and its staff to the virtual exclusion of any other commercial nuclear power issue. During the months immediately preceding the Great Bandwagon Market for light water reactors, in other words, the government's attention to nuclear power issues was largely confined to sorting out the technical and political pros and cons of the *next* power reactor technology. In August, the staff selected a sophisticated concept using helium as a coolant and graphite as a moderator, designed to operate at temperatures higher than any yet achieved. The "high temperature gas reactor" was, however, to be supported in a way which more or less paralleled the laissez-faire management philosophy which had controlled reactor development programs during the 1950s. The high temperature gas reactor program, as it took shape during the summer of 1964, was basically an industry program receiving government support. The series of events leading up to it was one in which the government played an essentially passive role, attempting only to balance competing technical claims and political pressures against its own internal priorities and financial constraints.

The assumption that government was in no position to question the technical judgment of private reactor manufacturers or their customers was in sharp contrast to the tradition from which Milton Shaw, the new reactor development director, came. Shaw came from Admiral Rickover's Naval Reactors Program. The cornerstone of the Rickover philosophy and, hence, in Shaw's view the key to the technical success of the submarine propulsion plants, was absolute control of industrial contractors by government engineers. The Naval Reactors Program was a "government program" in the sense that technically expert government administrators exercised daily control over the activities of their contractors.

Shaw was an ardent proponent of the Rickover philosophy, and was less than pleased with the gas reactor project. The reactor manufacturer and the electric utility company, partners in the proposal accepted by the Commission in 1964, were unable to agree upon contract terms between themselves. The utility withdrew from the project in early 1965. The reactor manufacturer, however, soon returned to the Commission with a new partner and plans for an even more advanced facility. Disagreement quickly broke out between Shaw and the Commission. Shaw wanted to revive yet another project—a sodium-cooled breeder reactor, called the "liquid metal fast breeder reactor"—as a government development effort modeled on the Naval Reactor Program.

He saw the gas reactor project as a continuation of old AEC reactor development projects. The government would, in effect, be gambling many millions of dollars without technical control of the outcome. Given the requirements of a breeder reactor development program, this would be a serious mistake.

Many on the Commission's technical staff believed that when the ultimate "customer" for a new technology was a private commercial interest—as distinct from the government itself, as was the case of the Navy reactors—it was the Commission's responsibility to carry its development only through the initial experimental phases and then turn it over to private industry. Even Shaw agreed with this distinction, especially to the degree that it implied that the initial commitment of $40 million to the high temperature gas reactor project was a "hard ceiling" and that the government would not bail out the manufacturer if it encountered difficulties. The problem with this project, as Shaw saw it, was the manufacturer's evident reluctance to accept full technical and financial responsibility. It was a "seriously underfunded" enterprise over whose eventual success the government would have little direct control, and it would divert resources from the breeder reactor program.[17]

The Commission, however, wanted the high temperature gas reactor prototype to proceed. Chairman Glenn Seaborg was particularly adamant about the need for the government to help develop an improvement on light water technology.[18] In addition, most of the members of the Commission believed that the President's Budget Bureau would insist upon "cooperative" reactor development projects in which the government's industrial partner accepted "the open end of the deal" and assumed ultimate responsibility for technical control. Shaw countered that even in extreme cases where a project's success was a matter of "life or death" to a company, the inevitable tendency in industry would be to "cut corners" as the magnitude and complexity of development problems became apparent.[19]

These arguments were the subject of much discussion, but they did not prove persuasive to a majority of the Commission. In April 1965 the Atomic Energy Commission decided to proceed with the high temperature gas reactor project over Shaw's reservations.

These basic issues soon resurfaced in connection with the role of the Argonne National Laboratory in Shaw's breeder reactor program. A new test reactor, known as the Fast Flux Test Facility, was a key part of that program. But for several years the Argonne laboratory had been promoting a much smaller but directly competitive project as part of its laboratory director's efforts to place Argonne in a position of technical leadership in breeder reactors. Moreover, there was considerable support within the industry for the Argonne conception of a breeder development program. Many people in the industry believed that projects like the Fast Flux Test Facility under Shaw's "dictatorial" control would divert federal money from cooperative government/industry development of a true prototype for breeder reactors.

The battle was joined during the summer and fall of 1965. Shaw argued that the Fast Flux Test Facility was essential to the resolution of a host of technical questions about the fuel for breeder reactors. These questions would have to be answered before an electric utility could sensibly make the huge financial commitments that construction of a breeder prototype required. It was, he claimed, the Commission's clear obligation to do for breeders what the Naval Reactors Program had done for light water reactors: provide the technical base for a prototype construction project—and in a government-controlled program.

Shaw won this battle. The Fast Flux Test Facility was approved in November 1965, and his victory signaled the first significant change in government reactor development policy in more than a decade. For the first time the government's program had become what the Democratic majority on the Joint Committee on Atomic Energy had wanted for years: a top-priority, government-controlled effort. Under Shaw, the Atomic Energy Commission assumed full responsibility for shepherding a new reactor technology from engineering concepts to full-scale prototype "demonstration." Moreover, in the following years, the AEC showed growing willingness to sacrifice other goals to meet this commitment.

Unfortunately, this bitterly fought change diverted managerial, financial, and political resources from the problems of light water reactors. Soon these problems would all but overwhelm the AEC's reactor regulatory program. But even more important, they would help destroy one necessary ingredient of a successful government reactor development effort which both Shaw and the Commission apparently overlooked: public confidence.

But none of this was yet apparent. Indeed, the coming years saw one of the seemingly great innovational triumphs of the century for American technology. During the late 1960s light water reactors became the commercially dominant nuclear power technology not only in the United States but throughout Western Europe and Japan.

CHAPTER 3

The Irresistible Force of Light Water

The 1960s: A Decade of Oil

CONTRARY to expectations, the 1960s were not a propitious time for the commercialization of nuclear power in Europe. Both coal and oil became more plentiful in the late 1950s, but the most dramatic developments affected petroleum supplies. Oil had been discovered in the African Sahara in 1954, and by 1959 it was flowing to the North African coast through the Bougie pipeline. The remaining trip to Western Europe was short and cheap. This development began what turned out to be nearly a decade of steadily falling oil prices and growing consumption, accompanied by major structural changes in the international oil business which also appeared to benefit Western Europe.

During the 1950s, the discovery of major new oil fields near the Persian Gulf and in North Africa combined with a number of technical advances to bring declining production costs and high earnings to the international petroleum industry. With complex pricing and production control policies, the large multinational companies were, for a time, able to prevent the new sources of supply from depressing prices. But, by the end of the decade, the stability of this system was threatened by the entry of the so-called "independents" into the market. The 1959 decision of the Eisenhower Administration to establish American import controls left the burgeoning Middle East crude supplies with nowhere to go but Western Europe and Japan. Oil prices in the chief European markets began to fall, and they continued to do so throughout the 1960s.[1]

Meanwhile, European governments made little effective progress in developing common energy policies. Thought and action seemed drowned by the veritable flood of cheap oil. By the end of the 1960s, the pattern of oil distribution in Europe had come to vary according to the self-interest of the multinational companies and the governments of each separate country. The only common ground among the latter was each one's reliance on oil imports; in 1970, the European Community produced less than 4 percent of the petroleum products it consumed.[2]

The principal consequence of the lack of indigenous oil resources was the creation of an enormous market for producers. Government control over this market varied considerably, from simple regulation in Germany to strong centralized control through the allocation of complex import quotas and the presence of relatively strong national oil companies in France and Italy.[3]

The American situation differed from that of Western Europe in two significant ways. First, due to the import control program, there was no equivalent to the European price decline. Second, the decade was one of sustained self-sufficiency in all energy resources, including oil. Since the war, the United States had steadily enjoyed cheaper energy than that available in Europe and Japan. When oil prices dropped sharply in Western Europe during the 1960s, this American advantage narrowed (though on the average, energy remained cheaper in the United States). The then declining price of oil was a strong incentive for many European utilities to switch from coal to oil for electricity generation.

Cheap imported oil had major consequences for European energy policies. First, it contributed to the decline of the coal sector in spite of aid since the mid-1950s from the governments of the coal-producing countries. Community authorities had anticipated coal consumption of more than 300 million tons by 1960 and more than 333 million tons by 1965. In fact, actual consumption reached only 240 million tons in the mid-1960s. Coal prices fell with those for oil, but coal was still unable to compete successfully with oil.[4]

Second, imported oil greatly hindered the growth of the nuclear power industry. In fact, the optimism of the 1955 Geneva Conference began to fade well before it became apparent that cheap imported oil was to dominate European energy policy throughout the decade. Five years later, many in the European nuclear community were obliged to recognize that they had seriously underestimated the technical difficulties of large-scale commercial nuclear power.

The British Response: Shift to an Advanced Gas-Cooled Reactor Design

English planners realized that their operating experiences from the 50 MW plants at Calder Hall provided only a narrow basis for extrapolating the construction costs of their new 275 MW "Dungeness" units. By the late 1950s it had also become clear to those influential in the British nuclear program that the technical difficulties of the nuclear fuel cycle had been considerably underestimated at the start of the decade. Unanticipated problems left them needing complex chemical and metallurgical capabilities, new materials for fuel rod cladding, and experimental data about the behavior of exotic materials at high temperature and radiation fluxes.

The British, who had the world's first significant nuclear power program, were also the first to recognize the magnitude of the difficulties of reactor commercialization. In October 1957, a fire broke out at one of the two plutonium production reactors at Windscale. Moderate amounts of fission products were released into the environment and during the ensuing several weeks milk products from surrounding farms for 500 square kilometers were forbidden for public consumption. For the first time, an accident at an operating reactor had consequences that were brought to public attention.[5]

A White Paper presented to the British Parliament in June 1960 by the Ministry of Power contended that the need for an immediate and sharp acceleration in the rate of installed nuclear capacity had passed because of "unforeseen" reductions in the cost of electricity from conventional power plants. The White Paper also noted that the cost of electricity from the first British nuclear power plants was "rather higher than earlier estimates." Also, the downward trend in the cost of electricity from coal was steeper than had been expected. It did not now seem likely to the Ministry of Power that nuclear generation would be cheaper than conventional sources until 1970. Accordingly, the government decided to slow nuclear construction. The objective now was to have 5,000 MW of capacity in operation by 1969.[6]

In April 1964, the British Government issued another White Paper on its nuclear program. This statement formally confirmed a decision which in effect had been made one year earlier: Magnox technology was to be abandoned in favor of a new design for a carbon dioxide-cooled system, the "advanced gas reactor." The Magnox fuel element imposed rigid limits within which development could take place.[7] Cost reductions were therefore largely achieved through modifications of engineering design—for example, use of prestressed concrete vessels which opened the path to construction of reactors of larger size—rather than improvements in nuclear technology.[8] The advanced gas reactor was seen as such an improvement. It offered the economically important promise of removing the severe temperature constraints of the Magnox fuel element, thereby allowing higher temperatures in the heat cycle.[9]

The advanced gas reactor was, however, to involve years of technical frustration and failure. But it is important to stress that the technical challenges were accepted and the frustrations endured chiefly because of the decisive importance given to economic considerations.

> ". . . the fact that the nuclear power plants which are at present [1962] being built will produce power more expensively than it could be generated in conventional plants built concurrently, has forced our engineers to aim at a higher rate of technological advance than is comfortable and would normally be considered wise."[10]

The French Response: Abandon Gas-Graphite for Light Water

In France, as in Great Britain, the initial response to the decline in the cost of fossil-fuel electricity was to increase the size of gas-graphite reactors and to attempt to improve their performance. The first French reactor connected to the national power grid at Chinon in 1963 had a generating capacity of 60 MW. In 1965 a 540 MW unit was ordered at Bugey, incorporating new core geometry and higher performance fuel elements. French nuclear engineers were initially confident that costs could be greatly reduced through design modifications.[11]

It soon became evident, however, that technical improvements could not keep pace with the continuing decline in the price of electricity from oil. In 1963, the "Commission PEON,"* a high-level government technical advisory group, claimed electricity from gas-graphite reactors was "just competitive" with fossil power at 7.50 mills per kwh. The twin 720 MW gas-graphite units planned some five years later for Fessenheim were expected to generate electriicty at an average cost of 6.4 mills/kwh, still about 15 percent greater than electricty produced by equivalent fossil plants.[12] In 1968, the committee concluded that gas-graphite reactors could be competitive with fossil plants only if new, high-performance fuel elements were developed.[13]

During the 1960s, then, French hopes for cheap electricity from nuclear power progressively faded. Simultaneously, enthusiasm about the technical possibilities of gas-graphite systems was replaced by disenchantment at senior management levels of Electricité de France, the government-owned electricity producer.[14]

Electricité de France had come to define its responsibilities in the broadest possible terms. By assuming total responsibility for all dams and later all thermal stations, the company had developed strong industrial capability. Confident in its technical competence, EDF wanted to tightly control all aspects of the nation's electrical program: planning, construction, and generation.[15] For many of the company's engineers, a major program of reactor construction was an obvious sequel to the now completed hydroelectric programs. The difficulty from Electricité de France's point of view was that, in the case of nuclear power, it would be obliged to deal with the Commissariat à l'Energie Atomique, the French Atomic Energy Commission, more or less on equal terms. That organization also had a broad view of its responsibilities, and

* Commission Consultative pour la Production d'Electricité d'Origine Nucléaire.

the inevitable rivalry over territory erupted into a crisis by the mid–1960s.

Electricité de France was the epitome of a French administrative agency. Its management brought a thoroughly rationalistic, technocratic approach to decision making. Marcel Boiteux, who had an established international reputation in econometrics, operations research, and decision theory, became general manager in 1967. He had been an early French proponent of such advanced management methods as nonlinear programming. Prior to his appointment, he had developed a sophisticated set of management practices for the agency based on statistical and numerical techniques. On the basis of his work Electricité de France was cited throughout the world as the electric utility with the most advanced analytic tools for demand growth planning and management.

Despite the early enthusiasm of company engineers for nuclear power and the successes of the gas-graphite program, senior management was cautious about nuclear power development. For Marcel Boiteux in particular, the declining price of oil dictated the correct posture. Nuclear power should be considered a backup for electricity generation, but so long as oil remained cheap large-scale nuclear development of the technology was unwise. As late as 1970, the senior management of Electricité de France still seemed unanimously confident that oil would remain cheap and that nuclear power was only a long-term solution for electricity supply in France.[16]

It was in this atmosphere that open warfare broke out between the Commissariat à l'Energie Atomique and Electricité de France over the control of nuclear power. Their battle soon became the center of a wider dispute on the entire industrial front in France between advocates of nationalized efforts in technological innovation and proponents of more open and competitive industrial development, even if that meant greater reliance on foreign technologies. The CEA-EDF fight provided ammunition to both camps.

The Commissariat à l'Energie Atomique's opinion about the development of nuclear power was straightforward: it had the basic responsibility for the entire endeavor. Electricité de France disagreed.

The French AEC had been created in 1945 as a research and development organization and, like its American counterpart, it had never developed any significant industrial experience or capability.[17] When the time came to build the gas-graphite prototypes, it had to give construction responsibility to several industries willing to work together on the venture. Electricité de France, in contrast, with its own industrial architects and engineers, saw no reason to abandon its practice of assuming complete control of all aspects of designing and building power plants. The ingredients for institutional jealousy could hardly have been more obvious. While the AEC's contractors were completing the last two plutonium-producing reactors at Marcoule, Electricité de France undertook design and construction of a new nuclear plant at Chinon on its own initiative. The Commission's responsibilities at Chinon were limited to design and delivery of the fuel. And although the Commission had selected a prestressed concrete vessel for the Marcoule reactor and advised the utility to do the same, Electricité de France followed the British practice of using a steel vessel for their power plant.

Government intervention had been necessary as long ago as 1955 to control the rivalry. At that time Electricité de France was given primary responsibility for building nuclear plants and the Commissariat à l'Energie Atomique was charged with providing the electricity producer pertinent technical expertise at various stages of construction. A permanent joint committee had even been established to coordinate the agencies' relations, but the compromise did not work especially well. Relations continued to be difficult at best during the late 1950s and early 1960s because neither ever fully accepted the putative role of the other.[18] The AEC was reluctant to accept Electricité de France's authority over the construction of power plants in the way that the utility's other industrial suppliers did. Under these circumstances Electricité de France would probably have preferred to deal with a private manufacturer rather than the Commission, which it saw as a major constraint on its own ability to control nuclear expansion in France.

This long-standing battle over administrative "turf" was complicated by strong philosophical differences about nuclear power. More concerned with economic considerations than the AEC, EDF continually recommended design changes to improve the system's performance. The AEC favored a far more conservative design approach. Moreover, while the AEC favored heavy water systems as a backup alternative to gas-graphite technology,[19] Electricité de France grew interested in American light water technology.

Again, on its own initiative, Electricité de France agreed with SENA, a Belgian utility, to build the first European pressurized water reactor at Chooz. The partners ordered the nuclear steam supply system for their plant in September 1961 from the Schneider Group, which owned the Westinghouse license through its subsidiary company, Framatome.

By 1965, Electricité de France was thoroughly dissatisfied with the AEC's attitudes and gas-graphite reactors. The utility accepted the British decision to abandon the Magnox system, which was very similar to the French gas-graphite machine, as proof that France would soon be relying on a reactor technology no longer pursued anywhere in the world.[20] More important, was it not true that Westinghouse and General Electric had realized spectacular successes with light water? The early turnkey deals and the economic promises they made in them were widely publicized in Europe and created the impression that a major commercial breakthrough had been made with nuclear power in the United States.

Electricité de France was receptive to the apparent evidence of American progress for two major reasons. First, it was eager to capitalize on its involvement in the Chooz project, which its officials believed was a model reactor construction because it was unencumbered by the participation of the Atomic Energy Commission. Second, Electricité de France enjoyed an especially close relationship with the Schneider Group, the Westinghouse pressurized water reactor licensee involved in the Chooz plant's construction. Indeed, M. Gaspard, head of the Schneider industrial group since 1963, had served 15 years as general manager and two as President of Electricité de France. He had long been a proponent of light water and his new responsibilities certainly contributed to the favorable impression Electricité de France had of that technology.

Severe technical problems developed at the 480 MW gas-graphite plant at Chinon a few weeks after its start-up in the fall of 1966. This trouble caused further acrimony. An investigation revealed malfunctions for which the manufacturers were held responsible. But news media commentary, partly inspired by the AEC, attributed them to the tight control Electricité de France customarily exercised over the manufacturers. With so little discretion, it was argued, the manufacturers did not have any incentive to do their job properly. This media campaign against Electricité de France showed the reactor manufacturers' dissatisfaction with EDF and their preference for the turnkey bids, as these were already currently practiced in most foreign markets.[21]

In December 1967, the government decided to build two new gas-graphite units at Fessenheim. Electricité de France allowed industry to bid for the entire reactor system. The industrial bid was prepared by Babcock & Wilcox, Schneider, and the Compagnie Générale d'Electricité. Not surprisingly, the quotation was considerably higher than the cost of any previous reactor,[22] for industry would be taking full responsibility for a reactor construction program for the first time. The high cost of the Fessenheim bid merely served Electricité de France's interests, however, because it appeared to demonstrate that gas-graphite systems could not compete economically with light water plants. On the pretext that costs were unacceptably high, Electricité de France rejected the Fessenheim gas-graphite bid and indefinitely postponed construction of the two new units.

The Atomic Energy Commission, however, refused to concede. According to the Commission, the technical problems with the Chinon plant were standard learning problems that were bound to occur in the early development of any complex technology. Moreover, the Commission believed that part of the difficulties were due to EDF management practices. The Commission challenged EDF's engineering competence and questioned the excessive control that EDF managers insisted on exercising over their industrial suppliers.

For Electricité de France—which in December 1967 had been allowed by the government to participate in the first large (900 MW) light water reactor built by the Schneider group at Tihange, Belgium—things were equally clear. The company's management did not share the Atomic Energy Commission's optimism about the technical status of the gas-graphite concept.[23] Moreover, it accused the Atomic Energy Commission of taking a dogmatic position on the relative merits of gas-graphite and light water reactors.

Within the French nuclear industry, there was no important proponent of gas-graphite technology. No single company owned a license for the entire system, as was the case for light water reactors; instead, know-how was fragmented among the two feuding government agencies and a number of industrial interests. Initially, several potential reactor manufacturers had favored gas-graphite systems because their development could take place in a protected market. But soon they realized that these reactors could not be exported easily. Further, it was now generally agreed in the nuclear industry that large capacity gas-graphite plants would be much more difficult to build than the more compact light water system. Basically this was due to the latter's use of enriched uranium fuel which allowed a more compact design.

In July 1968, an interministerial committee following EDF's position decided that the issue of the installation of gas-graphite systems at Fessenheim should be reexamined. This decision assured an EDF victory. Postponement of the reactors for Fessenheim signaled the effective abandonment of the gas-graphite technology, and virtually opened the road to light water reactors in France.

In 1968, Westinghouse increased its marketing effort in Western Europe and created two new subsidiary companies in Italy and Belgium. It was also moving into a strong position in the nuclear fuel sector, and in partnership with the French group, Pechiney-Ugine-Kuhlman, it was planning to build a fabrication unit in Belgium. Westinghouse refused any cooperation with the French Atomic Energy Commission on this project. Meanwhile, most European manufacturers had shifted to light water reactors. These developments all increased the pressure for a shift to American technology in France.

Within the Commissariat à l'Energie Atomique, proponents of light water systems also became more numerous. The successful start of the Pierrelatte plant, which produced highly enriched uranium for atomic weapons, had overcome a major technical difficulty in using light water systems to make electricity. With Pierrelatte, France had mastered the complex feat of uranium enrichment. So it was that by early 1969 the advocates of light water had greatly increased their pressure on the government. They argued that any further failure to adopt American technology was merely postponing the inevitable, that by ignoring the worldwide commitment to light water technology France was weakening its industrial position. Only by joining the light water club could French industry capitalize on the accumulated experience of American engineering and open the door for French export of nuclear technology.

The converging pressure from Electricité de France and French equipment manufacturers in favor of American technology was only reluctantly accepted in government circles. As President of the Republic, General de Gaulle had been only mildly interested in nuclear reactor programs; his chief nuclear objective was the creation of nuclear weapons. But a shift to light water reactors was hardly consistent with his nationalistic views on technological and economical development. However, his departure in April 1969 meant the removal of the last impediment to a complete French commitment to American technology.

In May 1969, the "PEON" Committee reexamined the question of nuclear power development and recommended to the government the construction of four or five 700–900 MW light water reactors on the grounds that they would produce electricity 15 percent cheaper than gas-graphite reactors of the same capacity.[24] Eventually, the government accepted the recommendation, allowing both American light water designs to be constructed. However, the Compagnie Générale d'Electricité, the General Electric French licensee, which had been the strongest advocate of the abandonment of gas-graphite technology, was ultimately disappointed. Framatome's bids were so low that the company received all of the new orders.

For the Atomic Energy Commission, the government decision meant the collapse of its hopes to continue commercialization of gas-graphite technology. The agency's senior management was bitter, believing that the decision was based on economic arguments which were questionable at best. But bitterness was hardly adequate solace, for the AEC was now excluded from further participation in commercial reactor development in France. Light water technology was entirely in the hands of French industry and its United States licensers. In the words of a well-informed observer, Michel Grenon:

There is some irony in the fact that, by deciding to develop light water reactors, France, whose (nuclear) program was strongly influenced by the desire for national independence, caused its industrial groups to be more tightly linked to their U.S. licensers than any other European manufacturers, such as Siemens which managed to disengage itself [from the license], or ASEA's Swedish company which independently developed its own light water reactor concept.[25]

These events in France in late 1969 were the last scenes in a drama that had begun more than 15 years earlier. After all the players had said their piece, the European nuclear power program had been dumped in favor of American light water technology. By the end of the 1960s, all Western European countries except England had chosen light water reactors as the basis of their nuclear power programs. In 1970, the "PEON" Committee summarized the situation:

In the various industrialized nations, the development of nuclear plants for electricity production has created a domain, the technical and economic importance of which increases steadily . . . Aside from Great Britain and Canada, the success of nuclear power is the success of American light water reactors . . . *which overwhelm all markets where true competition exists.* [emphasis added][26]

Few would have considered this assessment especially profound when it was made. It seemed to be merely official recognition of an altogether obvious situation: the almost irresistible market strength of American nuclear technology in Europe. The committee's judgment is interesting because it—and the conventional wisdom of the time which it reflected—is based upon a fundamental misunderstanding of the reasons for the commercial triumph of light water. The following chapters will elucidate this.

CHAPTER 4

Intoxication by Light Water

LIGHT WATER had begun to overwhelm all but the British and Canadian electric power markets by the early 1970s. In the United States, its success in competition with fossil-fuel alternatives was particularly striking. By the middle of the decade many informed predictions about American energy supplies for the last quarter of the century suggested a crucial role for light water reactors. Business and government planners saw nuclear power growing from a negligible share (less than 5 percent) of electrical capacity in the early 1970s to a considerable share (more than 20 percent) by the 1990s and a dominant share (about 40 percent) early in the 21st century. Similar projections were being made in many Western European countries and in Japan.

In the euphoria which accompanied this apparent innovative triumph, an important point attracted little attention: there was very little actual operating experience with large light water reactors. During the Great Bandwagon Market, government and industry planners believed that about six years would be needed to build the plants then being ordered. In practice it took much longer.

At the beginning of 1970, none of the plants ordered during

the Great Bandwagon Market was yet operating in the United States. Operating experience was limited to the nuclear power plants ordered prior to 1965. These represented an aggregate generating capacity of only 4,200 MW, compared with the 72,000 MW of capacity then on order or under construction.[1] The first light water reactor built without any direct government financial assistance did not begin operations until 1967.

Equally important, the economic characteristics of all the "pre-Great Bandwagon Market" plants bore little resemblance to the characteristics of those ordered in the mid– and late 1960s. This meant that virtually all of the economic information about the status of light water reactors in the early 1970s was based upon expectation rather than actual experience. The distinction between cost records and cost estimates may seem obvious, but apparently it eluded many in government and industry for years. Indeed, for the entire decade between the mid–1960s and the mid–1970s, few attempts to understand the economic status of light water reactor costs systematically distinguished fact from expectation.[2]

In the first half of this crucial 10-year period, the buyers of nuclear power plants had to accept, more or less on faith, the sellers' claims about the economic performance of their product. Meanwhile, each additional buyer was cited by the reactor manufacturers as proof of the soundness of these claims. By the early 1970s, the reactor manufacturers were pointing to the almost universal commitment of the American electric utility industry to nuclear power as a demonstration of its economic competitiveness.[3] For other potential buyers in the United States and abroad, as well as for those companies already committed to light water plants, this was a highly credible argument. The rush to nuclear power had become a self-sustaining process.

The gamble which the American manufacturers had taken in the early 1960s appeared to have paid off in a stunning success. By 1963 these companies had apparently decided that the time had come to capitalize on the experience they had accumulated as contractors for the government's military nuclear programs. The moment was at hand to translate this unique engineering know-how into success in the marketplace. The implicit assumption behind the turnkey reactor bids must have been that this engineering know-how reduced to acceptable levels the financial risks of fixed-price contracts for complete nuclear power plants which were much larger than any built so far.[4] The companies must also have assumed that the development of a fully commercial nuclear power technology would pose problems similar to those encountered in developing most new industrial products. Of course, both Westinghouse and General Electric sustained financial losses from the turnkey bids. Although the magnitude of these losses was reportedly much greater than anticipated, what seemed more important than the early financial setbacks was the ensuing strategic victory. The nine turnkey projects profoundly affected the American electric industry; their mere existence appeared to confirm the reality and imminence of commercial nuclear power. The Great Bandwagon Market followed.

But, from the reactor manufacturers' perspective, it was soon clear that things were not quite as good as they seemed. By the end of the 1960s, there was considerable evidence that the 1964–1965 cost estimates for light water plants had been very optimistic. The manufacturers themselves were prepared to admit this. But at the same time they contended that the causes of the first cost overruns were fully understood and were being dealt with. They were entirely confident that the combination of "learning" effects and engineering improvements in key reactor performance parameters (e.g., fuel life) could be relied upon to compensate for the unexpectedly high costs they were encountering. Economies of scale were also seen as a powerful tool for lowering the cost of electricity from nuclear power plants.[5]

By the late 1960s, light water manufacturers were in a headlong race to compete with conventional power by offering ever larger and larger plants for sale. The result is shown in Figure 4–1. Equally striking was the extrapolation of ordered plant capacity over operating experience. The dashed line in Figure 4–1 represents the rating of the largest commercial nuclear plants in operation for at least a year. By 1968, manufacturers were taking orders for plants six times larger than

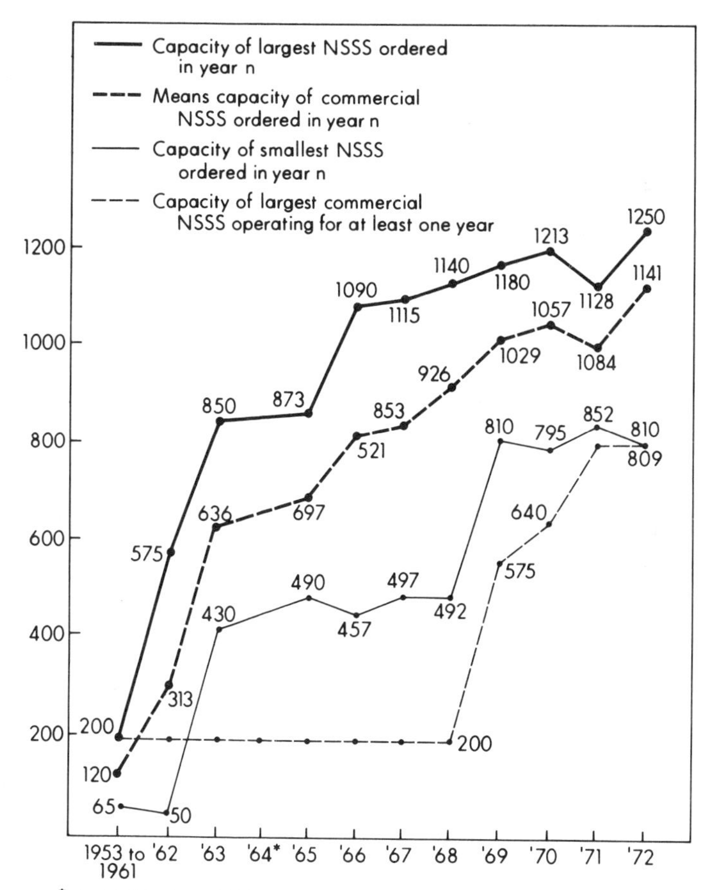

*No NSSS sold in 1964

Source: David L. Bodde, "Regulation and Technical Evolution: A Study of the Nuclear Steam Supply System and the Commercial Jet Engine," (Doctoral Thesis, Graduate School of Business Administration, Harvard University, 1975).

Figure 4–1

the largest one then in operation. And this was an industry which had previously operated on the belief that extrapolations of two to one over operating experience were at the outer boundary of acceptable risk.

Even Figure 4–1 understates the magnitude of the risks that reactor manufacturers and utilities accepted in the late 1960s. In the decade between 1962 and 1972, cumulative operating capacity—a measure of the depth of experience—became an ever smaller fraction of the cumulative capacity which had been ordered. Operational capacity declined from over 50 percent of ordered capacity in the early years of nuclear power to 3.5 percent by 1967.

Reactor manufacturers sometimes lost contracts for power plants by failing to keep pace with contemporary trends in plant capacity growth. Even though some companies suspected that size extrapolations would create problems, they had no way to prove their fears. The manufacturer of a smaller plant could not credibly offer better performance in terms of reliability and delivery schedule, so the larger plants that promised cheaper electricity won sales bids. The utility industry, for its part, did very little to restrain the manufacturers' rush to larger plants. In fact, the industry actively encouraged it by creating a bidding environment for plant sales which encouraged extravagant promises.

There has been considerable speculation about the motivation of the utilities which rushed to accept the turnkey contracts and so created the Great Bandwagon Market. The direct and indirect subsidies with which the AEC supported power reactor technology obviously helped make the new technology attractive to the electric utility industry.

A supplementary interpretation maintains that the investor-owned utilities feared they would be squeezed out of business by publicly managed companies which would exploit nuclear power in the same way that the Tennessee Valley Authority had exploited cheap coal to become a regionally dominant electricity producer. The management of many investor-owned companies was allegedly convinced that the only way to prevent the horror of "massive atomic TVAs" was to preempt the public utilities by getting a firm early grip on the generating technology of future nuclear power.[6]

A further, less obvious, but in our view very influential, factor was that the American utility industry was ready, even eager, for a new source of technological progress in the sixties. That industry was one of the great growth areas in the postwar American economy, enjoying favorable trends in almost everything affecting its profits. Ever since statistics for the industry had been kept, electric energy production had doubled every nine or ten years. But perhaps the most important characteristic of the industry during the years immediately preceding Jersey Central's decision was the *predictability* of its environment. In the decade preceding the Great Bandwagon Market, electricity demand grew at an annual rate of about 7 percent, with little variation. For individual companies, year-to-year variations in demand were larger, but they were the result of things such as population change and growth in households that utilities well understood. Forecasting growth in energy consumption was not considered to be a difficult problem in these circumstances, and so capital investment planning was easy.[7]

While demand was growing steadily and regularly, generating costs were declining: nationally, they dropped from a little more than 6 mills/kwh in 1960 to less than 5 mills/kwh by 1966. Most of this decrease was due to technological progress: economies of scale from even larger generating plants and improvements in operating efficiency.[8] The amount of "primary" energy from the combustion of coal or oil—or from atomic fission—needed to produce a unit of electricity is called the "heat rate," and after the war heat rates steadily improved.

But there were more personal reasons for the Great Bandwagon Market. Most potential buyers of light water reactors could look back on professional careers which had benefited from an unbroken series of technical advances; and to many of them nuclear power was an obvious continuation of the trend.

There were few, if any, credible challenges to this natural conclusion. Indeed, quite the contrary. Government officials regularly cited the nuclear industry's analyses of light water plants as proof of the success of their own research and development policies.[9] The industry, in turn, cited those same government statements as official confirmation of its own claims about the economic competitiveness of its product. The result was a circular flow of mutually reinforcing assertions that apparently intoxicated both parties and inhibited normal commercial skepticism about advertisements which purported to be analyses. As intoxication with promises about light water reactors grew during the late 1960s and crossed national and even ideological boundaries, the distinction between promotional prospectus and critical evaluation became progressively more obscure. For a dozen years after the Oyster Creek sale, and in at least as many different countries, favorable analyses, partly motivated by self-interest, were accepted as independently verified fact. Indeed what was missing during this entire period of intoxication was independent analysis of actual cost experience.

In the rush to light water, many lost sight of the fact that it was not becoming much easier to answer a simple but important question: "How much does a light water power plant cost?" In the heady years of the early 1970s, with the apparent worldwide triumph of American reactor technology, all that was really clear was that the cost of the plants sold during the Great Bandwagon Market would bear slight resemblance to the promises of their sales agreements. Even though more than 100 light water reactors were under construction or in operation in the United States by the end of 1975, their capital cost was almost anyone's guess.

It was widely advertised during the Great Bandwagon Market that light water plants could be built for less than

$150 per kilowatt.[10] For instance, Boston Edison's Pilgrim Station, ordered in August 1965, was expected to cost about $100 per kilowatt. Yet when it entered operation in July 1972, its final cost exceeded $300 per kilowatt. In 1974, Northeast Utilities estimated that its Millstone #3 plant ordered in 1966 and then scheduled for operation in early 1982 would cost more than $900 per kilowatt by the time it is completed in 1978. With this construction experience many in the utility industry began to suggest that light water resources entering operation in the mid–1980s might have capital costs of $1,000 or more per kilowatt. Do these estimates mean that light water reactors doubled in cost between 1965 and 1975 and will double again in less than 10 years? Not necessarily.

First of all, estimation is obviously complicated by inflation. A dollar spent in 1974 was "worth" less than a dollar spent in 1969; it bought less. Or, more technically, the purchasing power of "current" dollars varies over time. Because of this phenomenon, the cost comparison of nuclear power plants purchased at different times must be made in what economists call "constant" dollars. The difference between current and constant dollars is that the latter have been adjusted to take into account inflation and the former have not. For the rest of this book, all costs which have been corrected for inflation will be specifically cited as "constant dollars of year x." Assume that any data *not* cited in this way are in current dollars.

The problem of reducing capital costs to constant dollars is further complicated because reactors are built over a period of years. The cost record of any one plant is itself a hodgepodge of dollars of different purchasing power. The buyer of a hypothetical plant in 1967 might have paid for the nuclear heart—the so-called "nuclear steam supply system"—in 1968 dollars and the turbine generator in 1972 dollars. The total cost of any given power plant is usually reported as an aggregate of such expenditures, each expressed in varying current dollars. Without information about the time of various expenditures, it is difficult to convert cost records into a reliable constant-dollar figure.

An additional obstacle to understanding the cost variations of light water reactors is the very wide range in capital costs for plants entering operation in any given year, which is only partly attributable to geographical differences. For example, of the American plants which entered operation during 1973, the least expensive cost $213/kw and the most expensive ran to $413/kw—both in constant 1973 dollars.

The final difficulty is that, appearances to the contrary, actual cost records, as opposed to cost estimates, were scarce as late as the mid–1970s. By the end of 1975, more than 50 light water plants were completed and licensed for operation in the United States. But no plant ordered after 1968 was ready for operation, and only 3 of the 14 ordered during this year were complete. Moreover, only 21 of the 30 plants ordered in 1967 were in operation. Hence, as late as 1975, real cost data for light water reactors was limited to the plants ordered during the Great Bandwagon Market.

From the available cost records about changing light water capital costs, it is possible to show that on the average, plants that entered operation in 1975 were about three times more costly in constant dollars than the early commercial plants completed five years earlier.[11] But these early projects were, like Oyster Creek, turnkey plants whose reported cost to the consumer is well known to have been considerably less than their cost to the manufacturer.

The reactor manufacturers could hardly have been pleased by the losses they sustained on the turnkey contracts. But, the price adjustments which they made in the 1965–1966 period were supposed to take care of the problem. These adjustments were accepted without undue comment by both industry and government observers. The first large light water nuclear power plants were, after all, major technological innovations; and the costs of first-of-a-kind products are usually underestimated. It often requires some time to assess the real problems of a new technology and to adjust initial expectations to reality. Costs normally stabilize and often begin to decline fairly soon after a product's introduction. Throughout the period of the Great Bandwagon Market for light water, and in the years immediately following, the reactor manufacturers repeatedly assured their customers that this kind of cost stabilization was bound to occur with nuclear power plants.

But cost stabilization did not occur with light water reactors. On the average, the plants which began to produce electricity in 1975 cost about 50 percent more in constant dollars than the first *nonturnkey* plants which entered operation in the early 1970s. The learning that usually lowers initial costs has not generally occurred in the nuclear power business. Contrary to the industry's own oft-repeated claims that reactor costs were "soon going to stabilize" and that "learning by doing" would soon produce cost decreases, just the opposite happened. Even more important, cost estimates did not become more accurate with time. The magnitude of cost underestimation was as large for reactors ordered in the early 1970s as it had been for much earlier commercial sales. For example, in 1968, light water plants were expected to cost only about $180 per kilowatt. The actual average cost of the plants ordered in that year turned out to be about $430 per kilowatt—in constant 1973 dollars. On the average, the cost of all of the light water plants ordered in the mid– and late 1960s was underestimated by more than a factor of two in constant dollars. Moreover, this gap did not narrow between 1965 and 1970.[12]

Of course, none of this cost information was available to the electric utility companies who joined the rush to nuclear power in the late 1960s. All they could know with confidence was that light water plants had become far more expensive than anyone predicted during the Great Bandwagon Market.

But the reactor manufacturers *and* the Atomic Energy Commission constantly reassured them that everything was under control.

Some five years after the beginning of the Great Bandwagon Market, Philip Sporn was still the only prominent skeptic within the American energy utility industry. In 1969 he reported to the Joint Committee on Atomic Energy that:

> During the past two years there has taken place a remarkable and ominous retrogression in the economics of our nuclear technology. [The light water reactor] has lost position [in the competition with coal]. This makes it difficult to accept without something more than a grain of salt the statement of the AEC [that] "the outlook . . . for nuclear power continues to be very promising . . ."[13]

As usual, Sporn's views were not popular. The reaction of Westinghouse was notable, both for its lack of specificity and its cavalier dismissal of evidence which was contrary to the company's interests:

> . . . We were very disappointed in both the quality and the content of this latest in the series of reports Mr. Sporn has prepared for the committee on the economics of nuclear power. It lacks the coherence and relevancy of previous reports, and contains far too many unsupported and unsupportable assertions, observations, and conclusions. Its most glaring defect, however, is that it presents an assessment of the electric power industry in general and, nuclear power in particular, which is badly distorted and inaccurate.[14]

The Babcock and Wilcox comment on Sporn's 1969 report seems particularly casual.

> Mr. Sporn has, as usual, written a very complete report and has brought to light in an objective manner some of the problems facing the electric utility industry, the Government, and the public. Many of the questions he has raised are matters which relate to the needs and desires of the people of this country. We believe these will be extensively debated over the next several years and *do not feel that we, as a company, are qualified to discuss the matter.* [emphasis added][15]

Philip Sporn's lonely dissent from doctrinal orthodoxy was buried by an avalanche of material from government and industry which "proved" that early misestimates about the cost of electricity from nuclear power were things of the past.

Both government and industry shrugged off the repeated failure of nuclear power to match economic expectations as a fully understood and, hence, manageable aberration. Later, as the failures persisted, they dismissed them as somehow irrelevant because "the same thing is true for coal."[16]

In all the official, quasi-official, and private studies, reviews, symposia, conferences, hearings, and reports on the economics of light water reactors between 1965 and 1975, only Sporn expressed any sustained skepticism. Consider his reaction to what was perhaps the high-water mark for unrestrained optimism: the Tennessee Valley Authority's purchase of two 1,000 MW units—"Brown's Ferry 1 and 2"—at a total unit capital cost slightly *lower* than available coal-fired equipment. Sporn was harsh. He labeled the bids with which General Electric won the contract from TVA as "irrelevant to evaluations of the current competitive status of nuclear power. At the most charitable interpretation, the TVA bids obviously were designed to introduce nuclear power to the largest coal-burning region in the United States."[17]

But bad news continued to come, and by the beginning of the 1970s an intense debate had arisen in the United States about the comparative economic status of nuclear power. The National Coal Association repeatedly testified before Congress about "the debacle of the atom," claiming that "nuclear power had fallen on its competitive face, and that the bright promise of Oyster Creek had never materialized." It was possible, as it had been more than five years earlier, to argue that light water power plants were either "competitive," "almost competitive," or "not yet competitive." All anyone had to do was choose the assumptions and the data that supported his interest.[18]

Late in 1971, economist M. J. Whitman, of the United States Atomic Energy Committee, told the Fourth Geneva Conference on "The Peaceful Uses of Atomic Energy" that "the evolution in the costs of nuclear power . . . would under normal circumstances be classified as a traumatic, rather than a successful, experience. However, many of the trends which have affected the rise in the investment costs of nuclear plants have had similar effects on the alternative methods of generating power." Whitman concluded that "nuclear power remains highly competitive only because increases in fossil-fuel plants' investment costs and fuel costs have permitted it to do so."[19]

In the paper he presented at the Conference, Whitman explicitly accepted the American manufacturers' explanation about the reasons for past cost increases and the future certainty of cost stabilization and decline. His argument was that past cost difficulties represented a "prelearning experience"; future costs would inevitably decline because of learning effects. Whitman recognized that the cost of a nuclear plant delivered a decade after the first commercial sale in the United States would be effectively higher, by more than 200 percent. He labeled this a period of "intense learning." But he predicted that for plants coming into operation in the late 1970s, learning would begin to cause cost decreases. During the entire first half of the 1970s, this continued to be the conventional wisdom about the immediate economic prospects for light water reactors. It actually became the basis for a new surge of reactor orders; in 1972, American manufacturers sold 34 plants to domestic utilities. Literature from both the United States government and the electric utility industry reflected confidence that the causes of initial cost problems were now fully understood and resolved. Unfortunately, they were not.

European Naivité

This obfuscation was especially effective outside the United States. In Europe, the failure of the economic performance of light water reactors to match early promises was difficult to discern, in large part due to the marketing activities of their manufacturers. At the 1971 Geneva Conference on "The Peaceful Uses of Atomic Energy," the audience was presented with an American paper entitled, "U.S. Light Water Reactors: Present Status and Future Prospects."[20] According to its authors, the reasons for the worldwide commitment to light water reactors were their "design and operating simplicity, compactness, low cost, economical fuel cycle and demonstrated reliability and safety." The report noted that delays in licensing, manufacturing, and construction and from "many other evolutionary activities characteristic of a new industry" had been experienced during the first few years of commercial sales. But "*today many of these areas have stabilized and future developments are expected to be characteristic of an established industry.* Manufacturing and construction capabilities are now well developed and fuel cycle service has achieved a commercial basis. . . ." [emphasis added][21]

Ten persons were cited as co-authors of this report. There were two from each of the four American light water reactor manufacturers—Westinghouse, General Electric, Babcock & Wilcox, and Combustion Engineering; the remaining two co-authors were executives of Bechtel, the architect-engineering company with more than half of the light water reactor construction business. Bechtel's Kenneth W. Davis had for several years been the director of the Atomic Energy Commission's reactor development program.

Which areas had "stabilized"? Who "expected" future developments to be characteristic of an established industry? The fact is that this report was neither more nor less than what should be expected from a self-interested promoter; like other documents delivered to scores of other officially sponsored gatherings, it was an advertisement, not an analysis.

As a promotional document, the report was not especially objectionable, although it was no more informative than an advertisement in a trade journal. But it was objectionable to place such a report, which obscured the truth rather than illuminated it, in officially sanctioned and authoritatively sponsored technical literature. It was with self-serving documents like this one that the American nuclear industry created a persuasive illusion.

The advertisements were undeniably effective. Manufacturers in the fragmented European nuclear industry began intense competition to become licensees of the American companies. When the scramble to obtain light water reactor manufacturing rights was over, some 18 different European companies or consortia were offering light water systems for sale. Eleven were trying to sell pressurized water systems, and the other seven marketed boiling water systems. Most of the new merchants had little or no prior experience with light water plants and relied solely on the American licensers for technical and economic information.

The European situation was also characterized by the existence of protected national markets. The governments of Western Europe were eager to foster domestic industrial capability in atomic energy, and their interest led to the creation of several preferential markets for national manufacturers. This development was especially true in France and Germany, where effective competition was entirely limited to domestic reactor manufacturers. Inside their protected markets, very little competition was possible because of the manufacturers' lack of experience.

By January 1970, only three boiling water reactors and three pressurized water reactors with capacities greater than 100 megawatts had operated in Europe for more than a year, and even this fact somewhat overstates the extent of European experience. Operating experience with light water plants was spread across several countries. The most experienced of these, Germany, had operated three plants for more than a year: the boiling water prototypes at Grundemmingen and Linden (237 and 256 MW, respectively) and the pressurized water prototype at Obrigheim (32 MW). Italy had operated an earlier 160 MW boiling water plant at Garigliano and a 242 MW pressurized water plant at Trinoversellese. France had cooperated with Belgium only in the construction of the 283 MW pressurized water plant at Chooz.[22] This was the effective experience with commercial and demonstration systems on which Europe based its commitment to light water. Moreover, because all these prototypes had been built under different industrial arrangements, not all of the experience gained in one plant was directly pertinent to the construction of the others.[23]

The narrow experience with light water systems was particularly striking in the case of the French manufacturer, Creusot-Loire.[24] By the late 1960s this company's only practical experience with American technology had come from its limited participation in the Chooz program: delivering the pressure vessel, the pressurizer, primary piping, and a few other pieces of equipment.[25] French participation in the construction of the 870 MW Tihange plant was somewhat broader; this project, however, was not begun until late in 1969.[26]

This vacuum was a congenial environment for the American manufacturers. Their European licensees were too inexperienced to be skeptical. In fact, it was also to the Europeans' advantage to make expansive claims about light water performance to induce their governments to nurture and protect a domestic nuclear industry capable eventually of competing with worldwide exports.

In the United States, various government bureaus served as advertising agencies for the American light water reactor manufacturers. Their European licensees continued the chain on the eastern side of the Atlantic. The result was another link in a circular chain of mutually reinforcing claims based upon little experience but much hope.

The promotion of light water affected European governments as much as the European nuclear industry. Again, the example of France is instructive. In May 1969, light water cost information provided to the "PEON" Committee played a crucial role in settling the French government's five-year-old debate on the relative merits of light water and gas-graphite systems.[27] The information—provided to the governments by the French light water licensee, Creusot-Loire—combined Westinghouse advertising claims with the company's own candid admission of its ambition to establish itself as a light water exporter. The Westinghouse economic claims were, of course, expectations with no relation to actual costs for the plants recently ordered.

In its recommendations to the French government, the "PEON" Committee stated that enriched uranium light water reactor systems could be built for approximately $200 per kilowatt, while current gas-graphite systems were being quoted at more than $250 per kilowatt. Moreover, the Committee set operating costs for light water reactor systems at 2.66 mills per kilowatt-hour, compared with 2.70 mills per kilowatt-hour for gas-graphite systems.[28] The combination of construction and operating-cost advantages for light water plants meant a predicted total cost advantage of about 15 percent in favor of the American technology. The Committee did recognize, however, that this advantage would be obtained only if enriched uranium was available from the American government at current low prices.

The actual data made available to the "PEON" Committee are summarized in Table 1. One example can illustrate some of the problems with these data. The Committee quoted the Diablo Canyon plant's construction cost as $146 per kilowatt, but the plant turned out to cost about $500 per kilowatt. In fact, even after correction for inflation, the average discrepancy between the expected cost of light water reactors ordered in the United States in 1967 and their actual cost was greater than 100 percent. The cost projection made for the "PEON" Committee by what it called the "pessimistic" Boston Edison Company turned out to be about equally mistaken.

Cost information provided by German utilities to the "PEON" Committee from the Stade and Wurgassen projects might have been considered somewhat better because of the greater experience of German manufacturers, relative to those in France, with light water reactors. However, the interests at stake were identical on both sides of the Rhine.

The same game was being played in almost all other European countries. On the basis of economic analysis which originated in the United States, it appears that European light water licensees adjusted nuclear pricing to what was seen as necessary to ensure a market share in the new business. They simply duplicated the strategy American reactor manufacturers used several years earlier. The "PEON" Committee's conclusion that "the American light water reactor . . . overwhelms all markets where true competition exists," was a seriously misleading assessment of the reasons for light water dominance. It disregarded self-interest, nationalistic concerns, and indicated the abiding willingness to accept self-serving promotional tracts as serious, independent analyses.

During this whole crucial period there was no independent attempt within the French government to check, or even to seriously question, the advertised economic prospects of light water reactors. In light of the developments in the United States we have already reviewed, the supposed 15 percent cost advantage over gas-graphite systems seems tiny. It is far less clear today than it might have seemed at that time that light water reactors really did have a clear economic advantage.

Elsewhere in the European community, there was some recognition that the economics of light water were not en-

TABLE 1

Independent Sources of Information on LWR Economic Status Reported to the PEON Committee, April 1969

Year	Source of Information	Cost $/kw
UNITED STATES		
1967	• Expected cost of Brown Ferry III (1300 MW)	132 $/kw
	• Expected cost of Diablo Canyon (1060 MW)	146 $/kw
	• Expected average cost of U.S. plants ordered in 1967	150 $/kw
1968	• Expected average cost of U.S. plants ordered in 1968	170 $/kw
	• Seaborg (Chairman of the AEC) estimate	170 $/kw
	• Staszeski (VP Boston Ed Co.) estimate of cost of reactors in operation in 1975	210 $/kw
GERMANY		
1967	• Expected cost of Stade (630 MW)	140 $/kw
1968	• Expected cost of Wurgassen (612 MW)	140 $/kw
1969	• Expected cost of Flessingue (400 MW)	174 $/kw

SOURCE: *Les Dossiers de l'Energie*, Vol. I, pp. 107–9. Published by the French Ministry for Industry and Research.

tirely straightforward. According to a report from the EEC Commission:

During the 1969–1970 period, eight power plants equipped with light water reactors were ordered in the Community. Seven lay within the power spectrum of 770 to 1150 megawatts. The other eight had a more modest rating of 450 megawatts. The specific capital cost [expressed in constant 1970 dollars] of these units, which will enter service between 1973 and 1975, varies between $140 and $260 per kilowatt. Undoubtedly, these costs have not been established from strictly comparable data, but this is only a very partial explanation. This wide divergence seems chiefly to stem from the absence of interpenetration of markets, from the particular situation in each of these markets, from the commercial policy practiced by each particular firm, and by the diverging industrial structure in the countries concerned.[29]

Aside from observations such as this, however, the absolute level of light water reactor costs was not seriously questioned in Europe at that time. Cost increases in the American market were already reported, but the origin of early underestimates was now considered fully explained. The reassuring explanations American manufacturers offered for the cost discrepancies were fully accepted. According to the EEC report:

> For power plants ordered during the first half decade of the 1970s, there is every reason to expect that at constant dollar values, there will be a stabilization of prices at the 1969 to 1970 level. During the end of the decade, prices should take a downward turn. *There is no doubt* that the above-mentioned factors causing the high increases in capital cost in recent years will gradually be brought under control and that numerous uncertainties will cease to exist. [emphasis added][30]

The real consequences of the illusion of low-cost nuclear energy created in the late 1960s did not become apparent in either the United States or in Western Europe until 1973, when the Organization of Petroleum Exporting Countries revolutionized world energy economics.

CHAPTER 5

High Tide for Light Water

WHEN Jersey Central Power & Light announced its purchase of the Oyster Creek plant, many people mistook the plant's economic promise for accomplished fact. General Electric's assertion that it could build a nuclear plant that would generate electricity more cheaply than conventional generating systems seemed to them proof that light water reactor technology had come of age. As the years passed, however, it became apparent that all was not what it had seemed to be from Jersey Central's economic analysis of the Oyster Creek project and from similar analyses by other utilities in the ensuing months.

In order to sell a nuclear plant to the Tennessee Valley Authority in 1965, General Electric had to show that it would produce electricity for less than 3.7 mills per kilowatt hour.[1] Some 10 years later, Northeast Utilities justified continuing construction of its partially completed Millstone #3 plant by asserting that it would produce power at about 35 mills per kilowatt hour.[2] Even corrected for inflation, the 1975 cost of electricity from the newer reactor represented more than a five-fold increase over costs in 1965. Still, Northwest Utilities made a strong case that Millstone #3 would save money for the utility's customers in Connecticut and Massachusetts.

The principal reason for the continuing economic attractiveness of nuclear power in spite of these enormous cost increases was the skyrocketing price of fossil fuels, led by petroleum. Prior to 1971, Arabian crude oil delivered to the United States went for about $2.00 per barrel, and domestic crude sold at a little more than $3.00 per barrel. The higher price for domestic oil was maintained with import restrictions and state production controls.

It has always been assumed that for plants of equivalent generating capacity, the construction costs for a nuclear plant would be higher than those for a fossil plant. It has, therefore, seemed evident from the beginning of the commercial nuclear era that the key to the economic success of the technology was to prevent a virtually certain operating cost advantage (i.e., fuel and maintenance costs) over fossil plants from being wiped out by higher capital (i.e., construction) costs. Figure 5–1 shows the relative contribution of these elements to the cost of electricity for both nuclear and coal plants at the beginning of the 1970s. While capital costs represented about 45 percent of total cost of a kwh in the case of coal, they represented 73 percent in the case of nuclear power. Conversely, nuclear fuel counted for only 21 percent of the cost of electricity compared to 52 percent for coal.

In late 1970, international crude oil prices began to increase; it was an increase that was at first gradual and then dramatic. In September, Libya negotiated a new arrangement with the oil companies holding exploration and production concessions within its territory. Libya's success caused other Middle East countries to seek comparable tax increases. The Tehran Agreement of February 1971 provided for five years of automatic increases in taxes, but Libya, once again dissatisfied, negotiated a new tax package in the Tripoli Agreement a month later. A succession of agreements related to currency escalation followed. Next came the so-called "Participation

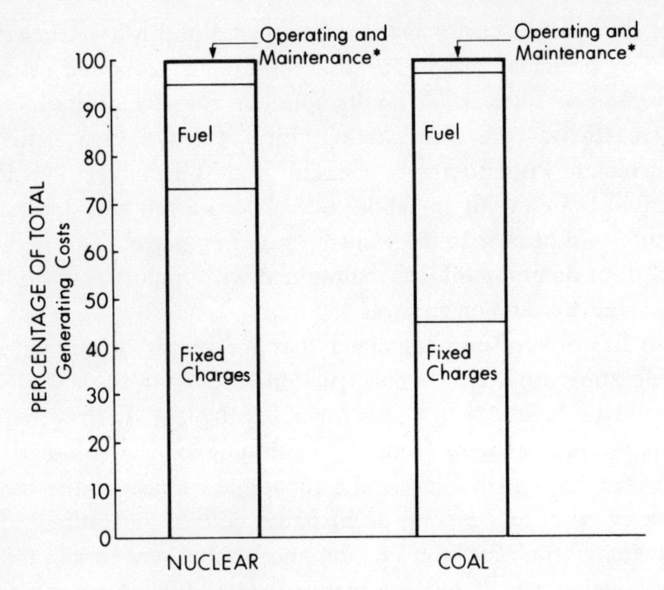

Figure 5–1

Agreements" of 1973, in which the oil-producing countries acquired partial ownership of their concessionary companies. In October of that year, the oil-producing countries in the Middle East reduced production and applied an embargo on oil delivery to the United States and the Netherlands in retaliation for their support of Israel. And two months later, the Organization of Petroleum Exporting Countries (OPEC) imposed a 130 percent increase on crude oil prices. Posted prices of Arabian light crude rose from $5.00 to $11.65 per barrel.[3]

During the previous decade, the coal delivered to American utilities had averaged about 25¢ per million btu. This price, too, began to slowly rise in 1969, and the rate of increase accelerated after the 1973 oil embargo. The "spot prices" paid by electric utilities without long-term delivery contracts almost tripled between June 1973 and November 1974, and the cost of coal supplied under long-term contracts almost doubled in the two years after 1975.[4]

At the time of the oil embargo, residual and distilled fuel oil were direct competitors with coal for generating electricity. In the United States, oil accounted for 15 percent of electric utility fuel and coal, 44 percent. The mix was more even on the East Coast, where imported oil was easily delivered and thus competitive with Appalachian coal. In fact, during 1966 many East Coast power plants began to shift from coal to residual fuel oil when import controls on it were, in effect, removed. They switched in even greater numbers with the Federal Clean Air Act of 1970, which restricted use of high sulphur fuels like most eastern coal.[5]

Thus, the effect of the 1973 oil embargo was quickly felt in the oil-dependent northeastern United States' utility market. Spot prices for residual fuel oil jumped almost 40 percent between October and November 1973. Coal prices soon followed. Spot coal prices rose about 10 percent in December 1973 and by an additional 30 percent in January. By March 1975, residual fuel oil had risen 282 percent above its price in June 1973. During the same month, the spot coal index rose 216 percent above its June 1973 level.[6]

Anticipation of a United Mine Workers' strike after November 1974 further increased the pressure on coal prices. Strenuous efforts late in 1974 by the steel industry, electric utilities, and foreign purchasers to build coal stockpiles caused a near panic that pushed up prices as production approached full capacity. Fortunately, the UMW strike was relatively short and the industry had resumed nearly normal production by January 1975. Spot coal prices then began a gradual decline. Long-term contract prices continued to rise, although more slowly than they had during 1974.

These two years of turmoil in the American fossil-fuel market were accepted by most informed observers in government and industry as having considerably altered the immediate prospects of nuclear power in the United States. By late 1974, the lingering squabble about the comparative cost of producing electricity from coal and uranium finally appeared to have taken a decisive turn. The nuclear industry mounted a persuasive campaign which stressed the long-term reliability of the nuclear fuel supply.

In the summer of 1973, Northeast Utilities released to the public a report on generating alternatives prepared by Arthur D. Little, Inc.[7] The consulting firm's principal conclusion was that ". . . the nuclear route . . . will yield the lowest busbar power costs and hence, from the consumer standpoint, has the most favorable economics."[8] For a nuclear plant ordered in 1973 to begin operation about 10 years late, capital cost of $700/kw still made nuclear power a bargain for Northeast Utilities.[9] The Arthur D. Little study claimed: "In absolute terms we project the cost [of oil] in the 10th year of plant operation (1991) to be about $2.30 per million btu [i.e., $12 bbl.]."[10] Two years later, the investment costs for new nuclear plants were being quoted at $800 to $1,000/kw. The price of oil, however, had already reached the level anticipated by Arthur D. Little in 1991. In a revised version of its 1973 study, the company was therefore able to claim that its original judgment about the relative economic merits of the two generating technologies had been confirmed.[11]

Figure 5–2 summarizes the evolution of the estimated cost of electricity (in constant dollars) from nuclear plants in the United States since the first reactor was connected to a distribution network in 1957. The estimate for any given year represents the advertised cost of power from nuclear plants ordered in that year. As we shall discover, this information allows us an interesting insight into the history of price competition between coal and nuclear power.

The cost of power from either of these two energy sources consists of three main components:

1. The capital cost (i.e., the fraction of total investment cost) to be charged to a kwh;

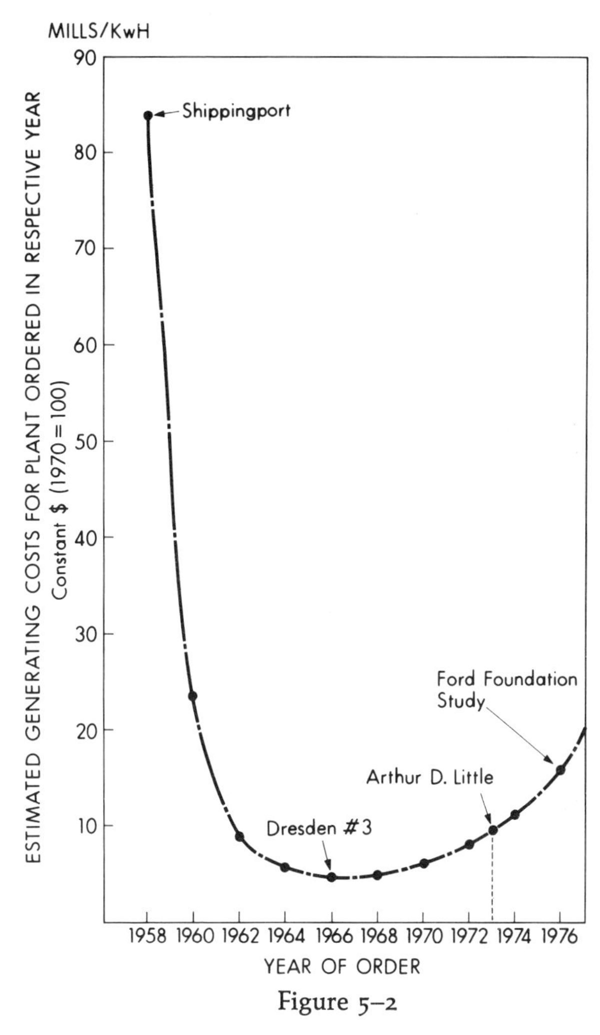

Figure 5–2

2. The cost of fuel necessary to generate that kwh;
3. The cost of operating and maintaining the plant during the production of the same kwh.

The nuclear power cost curve in Figure 5–2 is formed from the addition of these three components. A similar curve could have been constructed for the cost of coal-fired power, and we could have then compared the two. This form of direct comparison, however, is usually replaced by an indirect comparison, known as break-even analysis.

In break-even analysis, we are given the values of all three nuclear components and two of the three coal components, namely, the capital cost per kwh and the operating and maintenance cost per kwh. The purpose of the analysis is to cal-

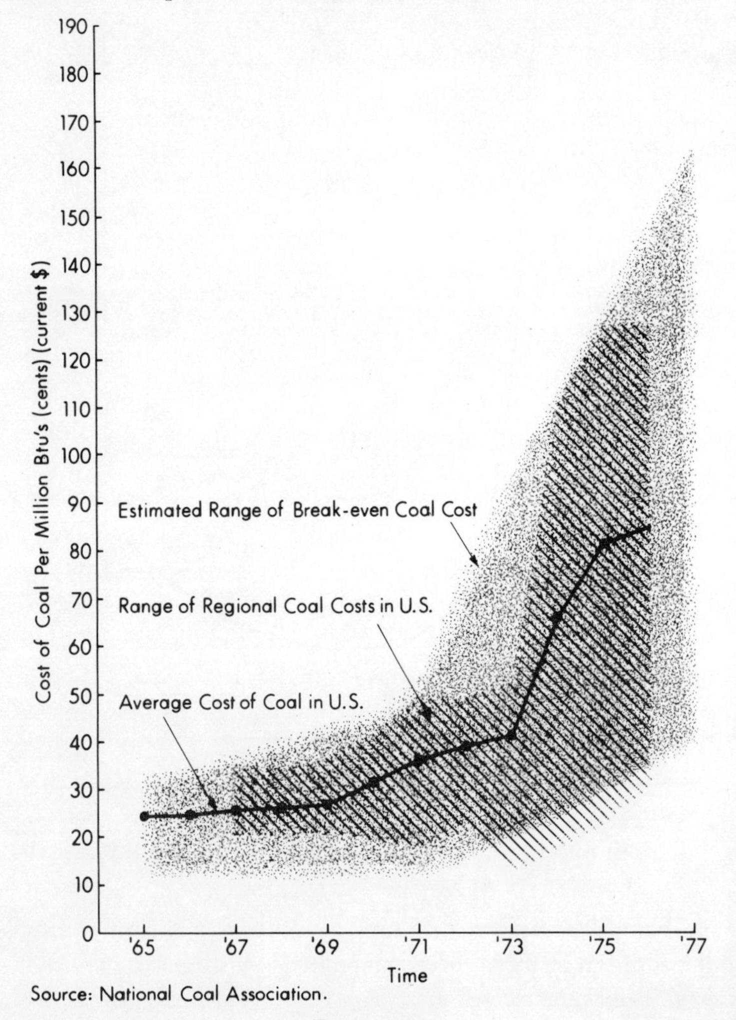

Source: National Coal Association.

Figure 5–3

culate the value of coal fuel costs which makes the total generating cost of coal-fired electricity equal to the total generating cost of nuclear-fired electricity. The "break-even coal price" represents the highest price a utility could afford to pay for coal and still generate electricity no more expensively than with nuclear power plants.

The trend of the hypothetical contemporary estimates of the break-even price of coal (Figure 5–3) is worth pondering. Coal seemed to be *just competitive* with nuclear power from light water reactors at about 25 to 30¢/mbtu in 1970; it still seemed to be *competitive* at about four times that price in 1976. On the same chart, the actual historic prices of bituminous coal in the United States have also been plotted (the hatched area). The correlation between the trend of the *estimated* break-even price range of coal and the *actual* market price is striking. Since the late 1960s it has only been because of steadily increasing coal prices that nuclear power could appear to remain economically attractive in the United States. Moreover, it is important to remember that the light water reactor generating costs on the basis of which those break-even costs were calculated were always *expected* not *actual* costs.

In the aftermath of the 1973 oil crisis, there must have been some relief in the nuclear power industry. The magnitude of the new increases in the prices of fossil fuels could be taken once and for all to have settled the economic issue. In addition, experts agreed that further increases for the traditional fuels were probable. After 30 years of doubt, nuclear power seemed, in a final and convincing manner, economically attractive.

During a conference at Harvard Business School in early 1975, Professor Paul Joskow of M.I.T. asked representatives of the United States light water reactor industry whether it was fair to conclude that OPEC had saved their business. The data we have just reviewed underline the pointedness of Joskow's question.

An Answer to OPEC: Light Water

By the beginning of the 1970s, oil had become the cornerstone of national energy policies in Western Europe. Many countries still believed that cheap energy from imported oil was their chief means of enhancing the attractiveness of their exports and promoting industrial growth. With the exception of the Netherlands and Norway, all of the countries of Western Europe were importing at least 50 percent of their total primary energy requirements on the eve of the OPEC embargo and price increases. Several, including France, Sweden, and Italy, were importing more than 75 percent of their needs. By the spring of 1975, the European reaction to the price increase was evident. If the 1960s had been a decade of imported oil, the 1980s would be a decade of nuclear power—nuclear power from American light water reactors.

In his opening statement to the European Nuclear Conference in April 1975, French Prime Minister Jacques Chirac stated: "For the immediate future, I mean for the coming ten years, nuclear energy is one of the main answers to our energy needs."[12] Similar objectives were assigned to nuclear power in other oil-importing countries. In December 1974, the European Economic Community announced a common energy policy. Reliance by the Common Market countries on foreign energy sources was to be reduced from about 60 percent in 1973 to between 40 and 50 percent by 1985, depending on oil price. Nuclear capacity was to double during the same period. In the United States, President Nixon's "Project Independence" articulated much the same commitment. Nixon proposed that atomic energy provide 30 to 40 percent of the nation's electrical generating capability within 10 or 15 years, and more than 50 percent by the 21st century.

In one sense, the rationale for these choices was simple and compelling. There were few realistic alternatives. The three possibilities for new energy were coal, conservation, and nuclear power. To have a realistic chance of affecting energy supply and demand patterns within 20 years, energy policies had to be based on some combination of these alternatives.

In the winter of 1973–1974, it was still possible for moderately well-informed persons to expect considerable development of shale oils, tar sands, and even the gases and liquids from coal before the 1990s. But within a year, such expectations were widely seen to be unrealistic. The accumulated evidence from the months following the OPEC embargo strongly supported the discouraging conclusion that no gaseous or liquid hydrocarbon substitutes for natural crude oil were available at costs much less than about $20 for the energy equivalent of a barrel of natural crude oil. Moreover, even if oil prices justified these costly alternatives, capital intensive and time-consuming construction programs would be needed before they could produce significant amounts of energy.

Among the three short-term alternatives to oil, the case for nuclear power was especially strong in Western Europe. The difficulties of expanding coal production were certainly real enough in the United States and even more pertinent to Europe: high cost, scarce labor, stagnant technology, environmental problems, transportation difficulties, and, for Europe, relatively limited supplies. At that time only a minority of energy planners believed that a strong energy conservation program should be a central feature of new energy policy. It seemed obvious to most political leaders that conservation on any significant scale would require changes in life-style that would be unacceptable to major segments of society and hence would be a political nightmare. And short of a drastic modification of everyday behavior, the immediate contribution energy conservation could make to a new supply policy was likely to be small.

With varying embellishments and varying degrees of sophistication, this argument very quickly became the premise of the national policy establishment's thinking in the United States[13] and Western Europe.[14] It was sound and reasonable. But few human activities are completely rational in the sense that the reasons which can be advanced in their support constitute a full causal explanation of them. This is especially true in interpreting the response of the leaders of the oil-importing countries to OPEC.

The development of nuclear power had been a consensus policy of the political, economic, and scientific establishments on both sides of the Atlantic for almost 30 years prior to the OPEC price increases. As a result, an imposing array of powerful and authoritative institutions in government, business, and academic life had become advocates of nuclear power. Their strength was especially visible in Western Europe. During the early 1970s, considerable concentration occurred in the Western European nuclear industry.[15] The strongest industrial groups to emerge from this process, Creusot-Loire in France and KWU in Germany, were eager to compete with the American nuclear industry for a share of the export prospects created by OPEC's actions.[16] The companies exerted considerable pressure on their governments to realize this ambition. The governments were more than prepared to respond favorably, since the companies' new ambitions perfectly matched the governments' own objectives: that nuclear power would not only be a major substitute for oil, but that massive exports of the new technology to developing countries would be a great help in compensating for the problems of balance of payments created by OPEC's oil prices.

The overriding purpose of nuclear advocates—to deliver the "benefits of the peaceful atom" to the world—had been legitimized by more than two decades of virtually unanimous support from the highest political leaders and some of the world's most respected academics.

In the heady months after the end of World War II, nuclear power had promised cheap and endlessly abundant energy. This promise was apparent to many of the most powerful scientific minds in Western society. To be sure, there were popularizers and science fiction writers eager to exaggerate and dramatize the promise with tales of "electricity too cheap to meter." The origin of this phrase is a minor mystery. But what is not mysterious is the source of the vision—that control of atomic energy was the crowning jewel to decades of intellectual exploration and triumph—which inspired it. It was equivalent in its practical implications to the control of fire itself. The source of this vision were the scientists who, even in the 1940s, were hailed as some of the most brilliant of the century. Their vision was quickly accorded similar honor and deference.

It is by no means an aspersion on the motivation and integrity of either the original atomic scientists or the younger scientists and engineers whose professional careers were later dedicated to the practical development of nuclear power to point out that by the 1970s they had become a powerful international pressure group. This group had available for Western political leaders what they were desperately looking for in 1974—a plausible response to OPEC.

The quality of the available economic information on light water reactors made the offer particularly hard to resist. During the preceding years the failure of economic performance to match economic promise had always been greeted by the nuclear community with confident assertions that things would be different tomorrow. These assertions, in turn, were accepted as proof that the situation was already under control. The political leaders who accepted nuclear power as their principal answer to OPEC had essentially no credible, independent economic evaluations of light water available to them. But what was even more important was that neither did many of them have very good information on the potential political costs of a commitment to light water reactors.

Since the end of World War II, suspicions and fears had,

of course, repeatedly been expressed about atomic energy. Many of the same scientists who originally saw the great promise of the new discovery also stressed its terrible dangers. For most of those who later worked to turn the promise into reality, these dangers, though real enough, were completely manageable. Convinced of this from their knowledge and experience, they tended to be impatient with outsiders who doubted it. But outside doubts never entirely died down. In fact, they actually seemed to grow in both number and intensity during the period of the early commercial successes of light water. But the nuclear community—and especially its leading scientists and engineers—retained the credibility to persuade the overwhelming majority of the public and virtually all business and government leaders that their rosy view was the correct one. Their serenity and confidence in answering questions about the safety of nuclear promise fully matched their certainty about its economic status. As the Great Bandwagon Market turned into an apparent worldwide commercial triumph for light water reactors, the trust of the political and economic establishment in the assurances of the nuclear community deepened. By the early 1970s it was easy to accept the judgment of the nuclear community that doubts were confined to a handful of vaguely informed or mischievously motivated persons.

This assessment of the situation was unfortunate. Policymakers missed an important political fact: the safety of nuclear power, indeed its fundamental social acceptability, was being increasingly questioned in almost all countries in which nuclear power plants were being built. This issue was becoming more, not less, troublesome at the time of OPEC's price increases. As others had done before them, the policymakers searching for an answer to OPEC accepted the nuclear community's claims about the political insignificance of the opposition to reactor technology. They had quite a surprise ahead.

Part II

THE EBB

CHAPTER 6

Reveille

IN EARLY 1974, the government of France announced its decision to shift the basic source of national energy from oil to uranium. This was the most important revision of French energy policy in two decades and came about between November 1973 and February 1974 with essentially no parliamentary or public discussion. The government considered it the only appropriate response to the conditions imposed by OPEC on France and other petroleum-importing countries. In its view accelerated construction of light water reactors was the only way France could regain some measure of energy independence.[1]

Within a year, the government's so-called "pari nucléaire" had become the object of more attention than any technical issue since the war. Between January and April 1975, virtually every major Paris daily and most mass circulation weeklies carried one or more articles on the subject in every issue. Questions were raised about many aspects of the government program. Both the safety and the cost of electricity from light water reactors were sharply questioned, as were the implications of relying on American technology. Attitudes ranged from denial that any problem existed to passionate assertion that some or all of the economic and political risks were unacceptable.

Political Background

Public discussion of such technical topics is highly unusual in France, which has a tradition of allowing the civil service to decide such issues with a minimum of outside interference. The plans and programs of technicians within government are customarily accepted with little comment. For two centuries, a powerful administrative apparatus managed by a highly trained technocratic elite has exercised responsibility for deciding technical questions according to its own perception of what was in the interest of French society. It has only rarely been influenced by public opinion.[2]

The power of the civil service technocrats is enhanced by several related characteristics of contemporary French political institutions. Members of the National Assembly are for the most part ill-equipped to deal with technical problems. French political parties, trade unions, and most elected politicians are by tradition chiefly concerned with social and economic matters. The existence of a large, publicly owned sector in the French economy is also relevant. Again, by tradition, left-wing political opposition and the trade union leadership have generally been confident that the technical choices made by publicly owned industrial and commercial enterprises are based on strict adherence to the "public interest." Even the most senior political leaders become directly concerned with technical issues only on the rare occasions where there is evidence of scandalously gross miscalculation or abuse of the public trust.

Highly technical national projects therefore took shape within the administration and the government through a process of confidential discussions among the agencies involved. The only opportunity for public debate occurred after basic decisions had been made and acted on.[3] But in the decades after the war, as the scope and effects of projects developed in this way grew, the ambiguity of the "public interest" become steadily more apparent. The result was a series of sharp but ineffective public debates: the Montparnasse "tower," the national computer program ("plan calcul"), the supersonic transport "Concorde," the slaughterhouse of La Villette. These disturbances suggested that the traditional decision-making process was coming under growing stress as it became more and more difficult to define the precise nature of the public interest in areas of advanced technology and industrial and urban planning.

In the area of electricity generation there were, however, few such concerns at the time of the government's decision to accelerate the light water reactor program. Electricité de France (EDF), had cultivated an extremely favorable public image. Of all the attempts to extend public control over areas of the economy after World War II—banking, automobile manufacture, air transport, railroads—the electricity generating agency was widely considered to be the most obvious success. In response to sharply growing demand, its engineers had designed and built sophisticated power plants for steadily declining costs, and French consumers were pleased to find their electric rates also declining.[4]

The Public Reacts Unexpectedly

Neither the government nor Electricité de France was prepared for the emotional public reaction which the 1974 light water reactor program provoked. At first, both assumed that ultra-leftists were simply making an issue of energy policy as part of their broader criticism of the French establishment. The government denied that there were any substantive risks associated with its reactor program and challenged the right of nontechnical observers even to discuss the matter. The electricity producer, for its part, maintained an air of serene confidence about its ability to cope with any problems that might arise during the proposed construction program.[5]

By late fall of 1974, however, it was clear that the government had badly mistaken the nature and strength of the opposition. Many groups at many levels of French society were discussing the alleged hazards of nuclear power: trade unions, local officials, citizen groups, and the scientific community itself. Less than nine months after making what it thought to be an entirely routine decision, the government found itself confronted with a burgeoning and almost unprecedented controversy about an issue previously considered the exclusive province of the technical establishment.

The activities of the second largest national trade union, the Confédération Française Démocratique du Travail, were particularly significant. Although the trade union officially favored a nuclear development program, its leadership questioned the need for the crash effort proposed by the government. It emphasized the risks and uncertainties possible in a rapid, large-scale nuclear expansion and called for a public debate on these issues. It also organized meetings across the country to inform its members about the content and implications of the government program. Strongly established among professionals in the French Atomic Energy Commission and Electricité de France, its scientists and technicians were perfectly able to understand and discuss technical aspects of the program. A group of these people prepared several reports on the potential hazards of nuclear power, the effect of which was magnified by a widely reported press conference in Paris in the fall of 1974.[6] After well-attended meetings in factories across France, the union's arguments were reproduced in widely read newspapers.[7]

The media were becoming daily more involved in the dispute. For several years Jean-Jacques Servan-Schreiber, the publisher of *L'Express*, had successfully cultivated the image of a responsible social critic. His growing interest in the nuclear program and the consequent attention it received in his widely read weekly significantly increased the respectability of the questions raised by the government's critics.

In early November, a report on the French energy situation was published by a special committee of the National Assembly.[8] Predictably, it completely endorsed the substance of the government's energy program. But it also criticized the government's handling of the problem, particularly with respect to public opinion. "Citizens are correct," the report stated, "not to accept on faith the claims of technicians who state that there is no danger and that all necessary precautions have been taken. . . . It is essential to permit everybody to understand and verify why there is no real danger." The report expressed concern that in the absence of such an understanding, ". . . misplaced fears and irrational apprehension could in the future delay and possibly cripple the entire French nuclear effort."

This warning reflected mounting alarm among officials that the situation was getting out of hand and that the government had to take some action to regain the initiative. Of special concern was the fact that only the limited question of how many light water reactors would be ordered in the coming two years had been decided. What would happen now that Electricité de France was beginning its investigation of possible plant sites? The search was undertaken in the traditional confidential manner, but the identity of several of the sites under consideration was leaked to the media, causing considerable unrest in some areas.[9] In order to ease tensions and prepare public opinion for a forthcoming additional commitment to nuclear power, the government decided to initiate a public information campaign.

The Government Campaign

On December 2, 1974 the Minister of Industry announced that the government would provide a full explanation of its reactor construction program. Moreover, the Regional Assemblies would be given the opportunity during 1975 to debate the location of all new nuclear power plants.[10] Following this announcement, the government sent to the regions a thick file on its nuclear plan. The file included a map noting the location of some 35 sites under consideration and preliminary information and data on each of them. The map was subsequently published by most daily newspapers.[11]

The government expected that these actions would demonstrate its willingness to openly debate the nature of the French commitment to nuclear power. It also expected that limited regional debate on the 35 potential sites would make it easier to identify those that would be the most acceptable ones. At no time, however, did the government suggest it might reconsider the decisions it had already made. It expected to quiet local opposition to plant construction, which, it assumed, was largely based on misinformation. The regional debates revealed a wide range of local attitudes toward nuclear plant construction. But the government did not succeed in appeasing its critics. The public clamor over reactor safety became louder in 1975.

In February 1975, the Conseil de Planification made public the energy program of the Seventh Plan: 12,000 MW of additional nuclear capacity would be ordered by the electricity producer in 1976 and 1977. This figure was 2,000 MW lower than had been proposed by Electricité de France and the Ministry of Industry. But, more important, as an ostensible concession to public concern about the implications of the nuclear program, the government also announced its willingness to have the program "formally debated" by the National Assembly in its forthcoming spring session.

Within a week of these announcements, a proclamation against the program was signed by 400 French scientists.[12] The proclamation asked "citizens to refuse to accept the installation of nuclear reactors until they have a clear understanding of the risks and consequences." During the same month a research group at the University of Grenoble publicized a study stressing alternatives to the government's "all nuclear" policy.[13] The audience for the report was considerable.

Locally, opponents of nuclear power—many identified with environmental protection groups—continued to organize meetings and demonstrations where power plants were being discussed. They received the cooperation of citizen groups concerned about the effects of reactor construction and operation on such activities as fishing and tourism. Between February and April, bitter controversy arose in a number of communities. In several instances, mayors responded to the conflicting pressures by holding referenda, the results of which varied widely. One small town in Normandy approved a potential reactor site by a two-thirds majority, while another on the Mediterranean coast rejected the proposal with equal decisiveness.[14]

These exercises of local democracy were irrelevant to national policy formulation. In the first instance, discussion of reactor sites had to take place in the Regional Assemblies, which were under no formal obligation to take local opinion into account. Created in 1972, the Assemblies had only limited power to participate with the central government in managing the social and economic life of their areas. Because they were not elected by direct votes, they were in practice little more than consultative bodies.[15]

The members of the Assemblies were almost uniformly embarrassed by the "nuclear files" given to them by the central government. To many of the "notables," or famous persons, who made up the Regional Assemblies, nuclear power meant the promise of additional employment and economic activity. Virtually all were completely unprepared to handle the technical and scientific issues involved in the nuclear debate. By March 1975, 10 of 22 Regional Assemblies had voted to accept the principle of building nuclear plants in the areas under their jurisdiction. Two had said no, and the remaining 10 had decided to postpone the decision. Most of them required further information before taking a final position on each site.[16]

The confusion of the Regional Assemblies was understandable. The debate over the government's "pari nucléaire" had become an increasingly strident confrontation between representatives of two extreme positions. A bewildering array of proclamations, statements, and demonstrations articulated conflicting points of view but rarely brought useful or even new information to the public's attention. Nor did either side indicate any inclination to compromise.

The leaders of the political parties were reluctant to become involved. Neither of the two main opposition parties—the Socialists and the Communists—took a formal position against the large-scale development of nuclear power in France. The Communists' opposition to the program was limited to concern about the government's preference for a foreign technology, the American pressurized water reactor. The Socialists' opposition was limited to technical aspects of the government's overall energy program; they advocated diversification, rather than concentration on a single source of energy supply, and an increase in energy conservation in order to allow the development of nuclear power "at a pace compatible with safety and energy requirements."[17]

Approval and Continuing Controversy

It was against this background that the National Assembly debated the government's energy program on May 14, 1975. It was, in fact, a very limited debate. It occurred one day after the President announced that the 1945 victory over Germany would no longer be publicly celebrated. This matter totally dominated the attention of the Assembly members, many of whom simply left the Chamber when the energy issue came to the floor. During the ensuing nine hours of discussion, not a single member seriously challenged the government's determination to rely heavily on nuclear power to meet France's future needs.[18] The formal nature of the proceedings surprised no one; the day before the debate began, a member of the governmental majority in the Parliament flatly stated in *Le Monde* that "France's energy choices have already been made."[19]

Protests and demonstrations against the program continued, sporadically, throughout the fall and winter of 1975. Meanwhile, EDF was preoccupied with the selection of the new plant sites needed to keep the construction program on schedule. The idea was to begin the licensing process in early 1976.

In many areas, however, local opposition prevented Electricité de France technicians from carrying out geological explorations. Company equipment was damaged at some locations, preventing further geological data gathering. In several instances, local authorities joined the population in demonstrations aimed at halting preliminary site surveys. By early 1976 there was a growing gap in several regions between the attitude of the Regional Assemblies and that of local political authorities. This gulf was particularly wide in Britanny, whose Regional Assembly approved of nuclear power but where Electricité de France faced mushrooming protests at every site it proposed.[20]

EDF's response to this situation was flexible. Wherever it met local opposition it postponed activity, continuing preliminary studies at sites where it encountered no serious challenges. The preliminary licensing process moved ahead at those quiet sites.

In early 1976, the Ministry of Industry took the initiative to start the administrative process for three sites. One them was located at Flamanville in Normandy, where a local referendum favorable to nuclear power had taken place the year before. The others were located at Cruas and Saint-Maurice-l'Exil on the Rhone River. Two new sites, one at Le Pellerin on the estuary of the Loire River and the other close to Thionville on the Moselle River, were added to the first three during the summer of 1976.

Not surprisingly, the beginning of the formal administrative process brought on new demonstrations by the nuclear opposition. At sites like Flamanville and Cruas, where the local authorities had advocated the nuclear plant, the conflict between opponents of nuclear power and the local population took on a new character. Sites that initially had been thought by the government to be politically favorable became the main focus of the national antinuclear groups. Increasingly prominent in these protests was a movement organized one year earlier by a group of scientists at the University of Orsay near Paris.[21] It had gathered 4,000 signatures of various members of the national scientific community against the government's nuclear program in the spring of 1976. The most active members of this movement had been taking a growing role in the regional public debates organized against the programs.

During the spring and summer of 1976, two sites were the object of particular attention by the nuclear critics. One was Nogent-sur-Seine. The relative proximity (100 km) of this site to Paris had made it an especially attractive target for the critics. Because of its location above Paris on the Seine River, nuclear critics charged that even a minor accident would threaten the Paris water supply.

The second was the site at Creys-Malville on the Rhone River above Lyon, chosen for the location of "Super Phénix," a prototype 1,200 MW sodium-cooled breeder reactor. The Creys-Malville project was a matter of particular concern for the nuclear critics because its reactor relied on an advanced technology. Creys-Malville was the principal focus of much of the attention of the nuclear critics through 1976 and 1977. An actual occupation of the site by the opponents was started in the summer of 1976 and the police forcefully evacuated the site.[22]

In late 1976 the imminent start of the Fessenheim reactor, the first French light water reactor to be started since the 1969 policy shift, also became the object of growing attention by local and national nuclear opponents. At issue was the emergency plan for the Fessenheim region in case of a nuclear incident. Critics contended that the population should be made aware of the details of evacuation plans before the plant began operating.[23]

In spite of the protests, the Fessenheim plant began operation in February, 1977. Anti-nuclear demonstrations continued, however, at Fessenheim and in other potential sites in Alsace.[24] Indeed, their frequency increased in all regions with nuclear plant construction projects, as did the apparent number of demonstrators at each one. There were, for example, 3,000 demonstrators at Le Pellerin in October, 1976; 5,000 at Flamanville in April, 1977; and 3,000 at Thionville in June, 1977.[25]

In spring 1977, a new "PEON" Committee was set up by the Ministry of Industry to make recommendations on the continuation of the nuclear program for the following two years. On June 27 the Ministry of Industry announced the government decision allowing Electricité de France to order an additional 10,000 MW nuclear capacity over the 1978–79 period.[26] Three days later, 30,000 demonstrators gathered at the Creys-Malville "Super Phénix" site to protest both the French breeder prototype and the government's nuclear program. About 100 persons were injured and one died during a clash between demonstrators and some 5,000 police officers protecting that construction site against intrusion.[27]

These two events represent the contrasting aspects of the nuclear power debate in France in 1977. A persistent and apparently growing nuclear opposition faced the government's continuing desire to move ahead with the nuclear construction program.

Between 1974 and 1977, for the first time, a debate had taken place on what would have been regarded a few years earlier as a purely technical matter that the national administration was capable of handling. The government's concession to allow public discussion of a technical matter was unprecedented. The technical judgment of public officials meeting in almost total confidentiality may no longer be automatically accepted as the basis for defining the national interest. In fact, the debate over the nuclear program may very well have changed the ground rules for the resolution of other issues as well.

For the moment, however, this debate has been settled by power. Except for the formal discussion of the government's nuclear file by the Regional Assemblies, the discussion of nuclear power essentially took place outside the channels of political decision making. It was more a confrontation than an actual debate.

The involvement of the political parties in the nuclear controversy has become increasingly stronger. In the 1977 spring local elections, nuclear power was an issue in several regions. Moreover for the first time in France, National Assembly candidates were presented by the ecology movement. To the surprise of many, they won more than 10 percent of the votes; in one of Paris' districts, they won 15 percent. And in most communities in the neighborhood of new nuclear sites and controlled by the left-wing coalition, local authorities openly opposed plant construction in their vicinity and the government program in general.

By late 1977 the level of passion in the nuclear controversies had not decreased. In October, 1977 the Socialist Party called for an 18-to-24 month moratorium on nuclear power plant orders and an immediate halt of construction work on Super Phénix. It meant that socialist success in the 1978 parliamentary election would probably lead to drastic modification of the French nuclear power program.

The 1974–77 nuclear debate was settled by power—but by power separate from authority. Electricité de France, for its part, was acutely aware of the precariousness of its position. A senior official privately assessed the chances of successful completion of the present nuclear program—a program he helped to develop and continued strongly to support—at less than one-third. At the same time, he believed there was an equal probability that political opposition would eventually be successful in seriously compromising the whole program. He was troubled that the outcome would probably be determined by passion, not by reason. In his view, French political institutions had not satisfactorily handled the nuclear power issue. The program was proceeding as planned, although with some delays, but with a heavy political mortgage.[28]

CHAPTER 7

Myth Against Reason?

PUBLIC OPPOSITION to nuclear power developed in France as it had earlier in other countries. The government and the affected business community also responded similarly. They proposed that the only problem was a distortion of certain well-established facts, and that serious consideration of the various technical questions raised by the critics of nuclear power was simply not justified. Given the correct information, they claimed, the public would quickly realize that there was no merit to the issues raised by an uninformed or even willfully perverse minority.

For most public officials and politicians, opposition to nuclear power was a temporary concern of a poorly informed fraction of the population. The nuclear critics exploited public ignorance and created the groundless impression among the public that many questions about nuclear safety remained unanswered. Clearly this impression would vanish as soon as the "true facts" were known.

In France it became popular among government officials to refer to nuclear power opposition as a *maladie de jeunesse* (childhood disease). Irrational opposition to novelty is presumably inevitable in the early stages of any technological innovation, they comforted themselves. So it became fashionable in establishment circles to compare opposition to nuclear power to the initial public distrust and fear of railroads in the 19th century. It was the responsibility of enlightened political leadership to guarantee the triumph of progressive rationality over obscurantist irrationality in such matters. The job required explaining to a misinformed public that there was really nothing extraordinarily novel, unknown, or dangerous about nuclear power; the obvious potential risks and hazards, while hardly negligible, were by no means unprecedented and were surely manageable.

These attitudes exhibited a startling misconception about the nature of the opposition to nuclear power. To many in Western society, science and technology are now identified with deterioration in the quality of life. Nuclear power has become for many of these people a powerful symbol of all that is wrong with science and technology. The cause of the nuclear safety controversy is a difference in values, not a lack of information.

It is, of course, true that nuclear power is hardly the first postwar technology to be identified with a degradation in the quality of life. And the fate of many of these threatening innovations suggests that such an identification need not necessarily slow down their development, especially if the technology promises economic benefits. But there are several reasons for thinking that the outcome of the nuclear power controversy may be different than the outcome of similar furors provoked by other innovations.

The first of these is the breadth and depth of the disagreement among scientists about apparently factual matters concerning the risks of nuclear power. Years of public debate on nuclear safety has clarified one essential point: no one agrees. Scientific unanimity about the risks of nuclear power is no nearer today than it was 10 years ago. For each of the technical questions in dispute, it is easier today than ever before to find apparently qualified scientists who give contradictory answers to many nuclear safety questions. In fact, since the early 1970s it has been virtually impossible to make any substantive statement about nuclear safety issues which was not challenged by either the proponents or the opponents as inaccurate or misleading.

The most important outstanding issues fall into four categories: the environmental hazards of normally operating nuclear plants; the likelihood and consequences of accidents; the hazards of the "nuclear fuel cycle"; and the chance of spreading possession of plutonium bombs.

Hazards from Normally Operating Plants

During normal operation, all nuclear power plants release minute but measurable quantities of radioactive material into the environment. Government standards in all countries stringently limit the release of such material, and strict procedures have been designed to enforce compliance with these regulations. In the United States some scientists have claimed that existing standards are too lax.

The nuclear industry is conceded to be correct in asserting that a transcontinental jet flight or a skiing vacation causes a person more radioactive exposure per unit of time than close proximity to an operating reactor. But the consequences of even a tiny incremental radiation exposure in various forms is not known with certainty. Some scientists have argued that the available evidence suggests that no amount of radiation exposure is completely harmless.[1]

In the United States, the overwhelming majority of technically qualified opinion seems content with existing federal regulatory standards. Informed criticism of nuclear power for low-level radiation exposure appears to be limited to a few scientists who disagree with the vast majority of their colleagues. But, resolution of the issue poses severe conceptual and practical difficulties.

The conceptual problem is that some versions of the challenge to present standards are in the form of a well-known logical trap: the impossibility of "proving a negative."

Suppose that I ask you to prove beyond a shadow of a doubt the following statement: "There is no way to turn lead into gold." Perhaps you spend years searching for a solution but to no avail. You still cannot claim to have proven the statement. In fact, you will never be able to do so. At any time, I can merely insist that you have not tried hard enough.

The same is true in trying to establish that minute amounts of radioactivity cause no harm. It will always be logically possible to claim that they might, but that we have not searched diligently enough to identify the damage.

The debate on low-level radiation effects reveals another characteristic of the reactor safety controversy. To most nuclear scientists and engineers the notion of "zero risk," which seems to be the standard many critics apply to reactor operation, makes no practical sense. They point out that all human activity entails certain risks which, presumably, are accepted in order to enjoy the benefits. The real problem is therefore balancing risks against benefits. To them the balance seems obvious for nuclear power. But, the even more acrimonious public debate about the likelihood and consequences of reactor accidents shows deep cleavage on both points between the advocates and the critics.

Reactor Accidents: Likelihood and Consequences

Two issues have been furiously argued. What is the probability of an accident to an operating reactor? And what would be the consequences of such an accident if it did occur?

Practically the only point on which there is agreement is that technicians must prevent certain circumstances from occurring. They must keep the nuclear core of an operating reactor constantly supplied with copious amounts of coolant to dissipate the heat produced by fission. In a light water reactor failure to do so for even a few minutes can cause the temperature of the fuel elements to increase beyond their melting point. Such a "fuel melt-down," like the stalling of an aircraft in flight, could initiate an irreversible chain of events, which for a reactor could end in the release of large amounts of radioactive material into the environment. There is, however, a wide difference of opinion between the nuclear industry and its critics about the consequences of the "worst imaginable" accident of this nature. In 1976 the American government published the results of a three-year, multimillion dollar reactor safety study. The Rasmussen Report estimated that the worst imaginable catastrophe—a loss of coolant, failure of all of the several backup systems, and fuel melt-down, during the worst possible weather conditions—would lead to some 3,300 fatalities, $14 billion in property damage, and require the evacuation of about 290 square miles of land.[2]

As serious as such damage and loss of life might seem, some of the nuclear critics allege that the Rasmussen Report drastically underestimates the consequences of the postulated accident. They have also claimed that the study was influenced by the strong desire of its financial sponsors—the Atomic Energy Commission—to minimize the hazards of a reactor accident. In 1977, American nuclear power critics published government correspondence which appeared to support this latter claim.[3]

The second contentious issue is the probability of such an accident. To estimate this, the Rasmussen Report used a statistical technique called "fault-free analysis," which was developed by the aerospace industry to predict the effect of failures of small components in large, complex systems. Thousands of possible sequences in reactor failure were assessed by computer for the probability of their occurrence. The combination of the resulting probabilities led to the prediction that the "maximum credible" reactor accident would occur once in every 10 million operating years. In other words, with 1,000 operating reactors, such an accident could be expected once every 10,000 years.

This probability, taken in combination with estimates of the maximum likely fatalities and damage, led the authors of the Rasmussen Report to conclude that nuclear power presents considerably less hazard to society than many other man-made or natural circumstances.[4] Some critics have refused to accept the validity of these conclusions.

Meanwhile, disagreement also persists on less basic issues. Everyone agrees about the need to incorporate emergency backup systems in all reactor designs as "second lines of defense." One such system, known as the "Emergency Core Cooling System" (ECCS), is designed to cope with a loss of primary coolant water in a light water reactor. The reliability of this hardware has been the object of an especially acrimonious controversy in the United States. The Union of Concerned Scientists, among others, has argued that there is a complete lack of experimental evidence that this system would perform properly in an emergency. The government, most academic scientists, and the nuclear industry insist that the lack is far from complete; full tests are an impossibility. Moreover, as further tests come in they show that previous standards were unnecessarily conservative. They reject the contention that it is irresponsible to allow reactors to operate in the absence of such evidence. Light water reactors in the United States today are presently permitted to operate under a set of more stringent "interim acceptance criteria" which were approved by the AEC after a lengthy public hearing.

Hazards of the Nuclear Fuel Cycle

The preparation and disposal of reactor fuel—the "fuel cycle"—consists of a sequence of industrial operations. Each involves the handling, conversion, and transportation of materials which in varying degrees are radioactive, chemically toxic, explosive, or all three. These operations are controlled by government standards.[5]

In contrast with the standards that control the construction and operation of reactors, fuel cycle standards, with some exceptions, have not until recently been the subject of public controversy. But at least two issues have now emerged in this area: radioactive waste disposal, and the protection of workers and the public from the toxic effects of the plutonium produced as a by-product of reactor operation.[6]

Certain elements generated in a nuclear plant remain highly radioactive and potentially dangerous for tens of thousands of years. No generally acceptable means of disposing of these materials yet exist. Indeed, the failure to aggressively pursue development of such techniques during the precommercial reactor development period is an embarrassment to the government agencies responsible for safety. Nuclear critics contend that present generations have no right to proceed with nuclear electricity production at what might turn out to be a very high cost to future society. The industry, on the other hand, is confident that finding an acceptable disposal method is simply a matter of time and money. The industry also contends—with government support—that the money and time available for research in this area are considerable compared to what informed technicians believe is necessary.

There has been much discussion of the relative advantages of two waste disposal strategies. One involves storing the unwanted material in closely guarded buildings designed to withstand any natural disaster or human activity. Another alternative is to put the material in a geologically stable but inaccessible place and seal it "forever" from human contact. Professor David Rose of the Massachusetts Institute of Technology points out that the two strategies are characterized by contrasting trade-offs. The first retains future options and provides greater assurance against error today, but it does so at a permanently high cost of administrative responsibility. The second approach, he notes, guards against irresponsible human behavior, but at a greater environmental risk. One strategy stresses retrievability; the other is designed for the opposite effect.[7]

In the United States no known human malignancy has ever been directly connected with plutonium, even though there are several well-documented cases of individual exposure to large amounts of the material.[8] But, there is experimental evidence that very small amounts (on the order of 10^{-6} grams) are fatal to animals when ingested into lung tissue in soluble form.[9] Two prominent American critics of nuclear power, John Gofman and Arthur Tamplin, conclude that this fact implies that a piece of plutonium the size of an orange would kill everyone on earth. Most other scientists strenuously reject this fearsome image. They argue that no data allow confident calculation of the lethal human dose of plutonium; in fact that there is no evidence at all about how much plutonium will start a cancer in humans. The nuclear power industry, in turn, cites these latter facts and scientific opinions to claim that plutonium is probably less of a hazard than experiments on animals would lead one to believe.

The Spectre of Nuclear Bombs from Plutonium

For 30 years, one of the chief obstacles to making nuclear explosives has been the scarcity of fissile material. A world energy supply system in which the role of nuclear power is growing necessarily implies that fissile material will become more plentiful because fission reactors produce much material as a by-product. Some highly publicized articles and books have pointed out that the plutonium produced by light water reactors is acceptable for nuclear explosives.[10] Only modest quantities of plutonium are needed for an explosive device. The cruder the device, the more you need. But, given some skill or luck on the part of the designer—and using reactor-grade plutonium to fashion a bomb requires relatively high skill—10 to 20 kg might be enough for a bomb that would pulverize the World Trade Center in New York.[11]

Today's light water reactors produce in a single year about 150 kg of plutonium that remain unburned in the core. The amount of material potentially available for weapons would become enormous if the nuclear industry grows at the rate planned during 1973–1974 by government and industry in the United States and abroad. Such growth means the annual production of several hundred thousand kg of plutonium per year by the turn of the century.

Critics note that by-product plutonium is subject to two threats. First is the problem of theft by criminals, political fanatics, and psychotics. Second is the possibility of seizure by legitimate governments.[12]

The first problem has recently become a prime concern of nuclear regulatory authorities in countries with existing or planned nuclear programs. The United States and Western European governments are currently strengthening the physical security of their nuclear power plants, fuel handling, and transport facilities. As is the case with the waste disposal problem, government and industry share the belief that satisfactory resolution of the security threat is only a matter of time and money.

Security measures to prevent the theft of fissile material by small groups and individuals by no means, however, solve all the problems of a world of abundant plutonium. The possibility remains that a government on whose territory reactors and fuel-processing facilities exist will simply seize fissionable material by *force majeure.* Nuclear critics point out that some countries which have announced plans to purchase reactors have not ratified the 1962 Nonproliferation Treaty. Critics and industry alike have argued that the only safeguards against a government decision to use reactor by-product plutonium for weapons purposes are diplomatic or, ultimately, military sanctions.

Hazards from normally operating nuclear power plants, the consequences of accidents, the dangers of nuclear weapons proliferation—these issues have fueled a stubborn controversy among scientists, government administrators, and business executives for more than two decades. But they are only one dimension of the reactor safety problem.

Further Reasons for the Persistent Nuclear Safety Controversy

Disagreement among scientists is only one characteristic of the nuclear safety controversy. There are other reasons to believe that opposition to nuclear power may be far more persistent than opposition to many past industrial innovations. The introduction of light water reactors was managed by government and industry in a way which enhanced rather than allayed doubts and fears, particularly in the United States.

During the critical early years of public visibility—the Great Bandwagon Market—the business and government interests with the most to gain from nuclear power's acceptance were the only people interested in information on the innovation *and* they monopolized the competence to assess it. The way in which these groups handled information about light water reactors and their impatience with questions from the outside surely contributed to the critics' sense that they were hiding something. By routinely impugning the competence or even the right of anyone "outside the club" to raise questions, government and industrial promoters of light water reactors enhanced the credibility and power of their critics. Antinuclear partisans were then able to create the widespread impression that much of the truth about the dangers of nuclear power was cynically concealed or distorted by government and industry. Once established, such impressions do not disappear quickly.

Yet another reason to suspect that the safety controversy will not be quickly settled is the heavily negative symbolism attached to nuclear technology. No economic or technical facts will easily erase the image of that mushroom cloud. Once, after listening to an enthusiast for nuclear power review the dangers to health and property of burning coal in great amounts, Harvard political scientist Professor Gary Orren responded: "Yes, but the fact remains that we did not drop a ton of coal on Hiroshima."

Critics of nuclear power use such images adroitly.[13] They repeatedly put government and industry officials on the defensive by asking such questions as whether plutonium is more toxic than botulism. To many curious citizens, the simple fact that such questions are discussed is message enough: whatever its precise magnitude, there is something enormously unpleasant about nuclear power. Talk of "unprecedentedly deadly" materials, "invisible" radiation killers, and lone psychotics fashioning atomic bombs may all seem melodramatic and irrelevant to the nuclear engineer. But it is powerful stuff. And the fears it arouses resist the rational calculus of comparative economic costs and benefits more strongly than most officials have supposed.

When a nuclear critic proposes that among all human attempts to change the natural order of things, nuclear power is the final and ultimate folly, we are not involved in testing a scientific proposition. Rather, we are confronted by superstitions almost as old as civilization. Nuclear power has caused the myth of Prometheus to rise to consciousness again. The myth is neither true nor false, nor is Weinberg's proposition that nuclear power is a "Faustian contract." But these beliefs demand society's attention.

They also reinforce yet another unhappy image linked to nuclear power: that of a small, elite group of scientists and technicians making decisions about what is good for society. This image collides with the growing demands of many in Western society for more participation in all matters which affect them. In the past, it was usually relatively easy for proponents of technological change to establish the desirability of their proposals. The fact that those who benefited from them were sometimes different from those who paid the price was usually overlooked. And in most cases there were few, if any, ways for the latter to air their grievances.

Much of this has changed during the past quarter century. One of the most important political developments in all of the Western democracies since World War II is that most have become far more representative. Many who were effectively disenfranchised throughout the Industrial Revolution now have the means to influence their societies. The ecology movement is one such group. By the middle seventies there were organized efforts in all of the Western political democracies in opposition to the further intrusion of industrial progress on nature. Much of the criticism of nuclear power was based on this point of view.

The political legitimacy of the nuclear critics is largely independent of the technical validity to their claims, even when they are supported by a consensus of established scientific and engineering belief. The opposition to nuclear power reveals a deep cleavage in Western society. Contrary to the assumptions of government and business leaders, this opposition is not likely to disappear when the facts are established and commonly understood. Resolution of the nuclear safety controversy depends less on whether the political and economic leadership of the industrial world is correct in its collective technical assessment of nuclear power than it does on whether this judgment is accepted as legitimate by their constituencies.

CHAPTER 8

The Imbroglio

THE centralized and elitist French political system showed itself to be ill-equipped to handle the intense public concern about nuclear power. Sweden, in contrast to France, has a long history of open public debate on a wide range of national policy. Its open decision-making process allows the Riksdag, the national parliament, to become a forum for considering issues routinely handled with varying degrees of administrative confidentiality in other societies. In many ways the Swedish democracy can be considered as an extreme contrast to the French technocratic structure.

In the United States, such issues as reactor safety are discussed as openly as in Sweden. But here, debates are far more disorganized and are not usually linked to formal party politics. For this reason, American critics of nuclear power have resorted to the judicial process rather than the political process to press their case.

A comparison of the course of the safety debate in these two countries will illuminate how very different political institutions have dealt with this issue.

The Nuclear Safety Controversy in the United States

The controversy over reactor safety in the United States has been an especially complicated one with many turning points. But, in general, it has evolved through three phases. The first phase, which coincided with the initial round of commercial orders for nuclear plants in the mid–1960s, was characterized by scattered, and largely local, opposition to specific projects. Often the object of concern at this stage was "thermal pollution." Light water reactors operate at a thermal efficiency rate about 18 percent lower than that of fossil plants: they release a relatively large amount of heat into the environment. In the mid-1960s the law did not require utilities to take any special measures to deal with this "waste heat," so it was common practice to disperse it in rivers, lakes, or the ocean. Since many forms of aquatic life are sensitive to temperature changes, this practice became a source of intense concern to environmentalists and, on occasion, commercial fishing interests. During this phase, opposition in a few instances caused outright cancellation of the project; in others, opponents managed to require substantial modification of the original plant design.[1]

This period culminated in the famous Calvert Cliffs decision. Calvert Cliffs was the local name for the proposed site of a nuclear plant in Maryland. On this occasion, nuclear critics challenged the Atomic Energy Commission's licensing of the project, contending that it had not considered potential environmental consequences of thermal pollution. The court supported the challenge, and the effect of its decision was dramatic. Reactor licensing came to a standstill for 18 months while the regulatory agency restructured its licensing process. A less immediate consequence eventually proved to be of greater significance: a series of supporting judgments in the federal courts followed and produced a body of case law that tipped the balance of legal power from the utilities to their critics.[2]

The second phase of the reactor safety controversy in the United States revolved around the alleged inadequacy of the system designed to cope with a loss of coolant in pressurized water reactors—the "emergency core cooling system." In 1966 the Atomic Energy Commission formed a task force of scientists and engineers from its own staff and from the reactor manufacturers to consider safety problems in building light water power plants of 1,000 MW, some 10 times the average capacity of the earliest designs. The task force reported that more research was needed in certain areas, especially "emergency core cooling," in the event of a loss of coolant.[3]

Although the Commission responded by redesigning a long-planned investigation of core meltdown into a test of the emergency cooling system, very little pertinent new information was developed in the ensuing four years. In 1971, experiments were carried out on an apparatus designed to model the behavior of a pressurized water reactor that had lost its coolant. The result of these experiments differed considerably from the predictions of a pretest computer simulation. The discrepancy was a problem and an embarrassment to the AEC. Although it was clear that the test apparatus itself was an inadequate model of a commercial pressurized water reactor, discrepancy between test results and predictions implied that there might be basic flaws in the calculation techniques then in use for evaluating the performance of the reactor core and its emergency cooling system. Hence, these limited experiments enhanced rather than reduced uncertainty. They made it possible to maintain that the computer programs on which commercial safety standards were based were inadequate.[4]

The Atomic Energy Commission responded to the situation by revising its standards for Emergency Core Cooling Systems. On June 29, 1971, the new and presumably more conservative standards—"Interim Acceptance Criteria"—were issued for immediate enforcement. Critics quickly challenged their adequacy. To deal with their complaints, the AEC held a public "Rule-Making Hearing" which convened in January 1972 and lasted until July 1973. In some 125 days of testimony, more than 22,000 pages of transcript and over 1,000 supporting documents were produced. The government issued revised "Emergency Core Cooling System Interim Acceptance Criteria" in December 1973, but again over the opposition's objections.[5]

The "Rule-Making Hearing" was a watershed in the history of the American nuclear safety controversy. There had been little prior public exploration of reactor safety policy and the manner in which the AEC made decisions. The inherent drama of the issue—the chances of a major catastrophe—and the colorful personalities of the participants focused Americans' attention for the first time on the question of whether nuclear power was safe. Prior to the Emergency Core Cooling System debate, opposition to nuclear power in the United States had been essentially limited to local challenges to specific power plant projects. In the aftermath of this debate, the issue became the overall social acceptability of reactors.

The Calvert Cliffs decision and the case law which it produced gave the nuclear opposition legal firepower. The Emergency Core Cooling System controversy added technical credibility to their case. It did so in part by publicly revealing disagreements between the AEC's research scientists and its regulatory staff. Prior to the hearings, American critics of nuclear power had been confronted with a monolithic opposition which monopolized the technical expertise on important aspects of reactor safety. Since the hearings, the Union of Concerned Scientists and other critics have been able to buttress their own claims with opinions from scientists and technicians from within the nuclear community.[6]

In the wake of the dramatic and nationally publicized Emergency Core Cooling System hearings, opposition to nuclear power in the United States entered its third phase by appearing to assume some of the characteristics of a national movement. Many would date the emergence of such a movement from November 1974. During this month more than 90 members of such organizations as the Friends of the Earth, the Union of Concerned Scientists, the Committee for Nuclear Responsibility, and the National Intervenors attended a conference in Washington, D.C. Organized and sponsored by Ralph Nader's Citizen Action Group, the conference was called "Critical Mass '74." Its objective was "to provide a national focus on the risks and consequences of nuclear fission and expand the citizen base—to move toward a nuclear moratorium."

For the first time, the American nuclear safety controversy was introduced into the country's formal political institutions. In the summer of 1975, the formation of a new national organization called "Critical Mass" was announced. Its sponsors claimed hundreds of affiliates and more than 200,000 supporters. Both Critical Mass and the Union of Concerned Scientists were national organizations in the sense that they drew financial support from a broad base of institutions and individuals. The American antinuclear movement was not, however, national in the sense of having a central organization controlling and directing antinuclear activities throughout the United States. It was still quite different from, say, the lobby against gun control, The National Rifle Association. Indeed, the antinuclear movement was still little more than a loose coalition of variously funded and variously motivated groups and individuals.

But by late 1975 it had gained a new and valuable ally. The prevention of proliferation of nuclear weapons' capabilities has been a constant goal of American foreign policy since 1945, and a loosely defined group of intellectuals and public officials had become professionally committed to the attainment of this goal. During 1975, this "arms control community" became increasingly concerned about the implications of light water reactor sales abroad. A number of well-publicized books and articles appeared and they explained that plutonium produced in light water reactors could be made into bombs.[7]

On August 6, 1975—the 30th anniversary of the bombing of Hiroshima—23,000 scientists, engineers, and physicians signed a statement urging a slowdown in the construction of nuclear plants and calling for the development of alternative energy sources. Among them were several prominent members of the intellectual and scientific establishment not hitherto associated with the antinuclear position, including James B. Conant, George B. Kistiakowsky, and Victor Weisskopf. The mere association of prominent scholars once largely indifferent to nuclear safety issues with the antinuclear movement was remarkable: the intellectual establishment had once been unanimously and unguardedly optimistic about nuclear power.

Having progressively enhanced the credibility of their claims and broadened their national constituency, nuclear opponents decided the time had come to take their case to the voters. A statewide referendum in California in June 1976 was the first of several attempts to secure voter approval for a moratorium on nuclear plant construction in the state. Given their momentum, it may be surprising that all of the critics' referendum efforts were unsuccessful. The course of the Swedish debate is revealing about the reasons for this setback and its probable consequences for the American antinuclear movement.

The Nuclear Safety Controversy in Sweden

In Sweden, as in the other oil-importing countries, the formation and early success of OPEC caused business executives and public officials to reassess their nation's energy picture. The Swedish government appointed several committees to make inventories of available and potential energy sources, projections of potential needs, and surveys of pertinent research and development programs. At the time of the first large price increases, Sweden was dependent for almost 75 percent of its total primary energy on imported oil. Another 15 percent of its energy supplies came from hydroelectric power, and although several large rivers remained, an aggressive opposition to their exploitation had developed in the 1960s. Sweden had no known coal or natural gas reserves. Even before the 1975 OPEC embargo, the issues of energy planning in Sweden had crystallized around the question of nuclear energy. Sweden was one of the few oil-importing countries that had both substantial deposits of uranium and a highly developed nuclear technology.

The organization responsible for the coordination of electricity supply was the "Centrala Driftsledningen" (CDL), or "Central Power Supply Management." By 1972, an ambitious reactor construction program was the basis of CDL's plans for expanding production. CDL did need the consent of the national parliament (the Riksdag) to finance this construction, but throughout the 1950s and 1960s all Swedish political parties had almost unanimously favored "the peaceful uses of atomic energy." In 1970 and 1971, Riksdag resolutions calling for the construction of the country's first 11 nuclear reactors were approved unanimously.[8]

A 1972 CDL report on electricity supply projected a need for 13 more nuclear plants between 1975 and 1990, and these plans had the strong support of the industrial and government establishment. In announcing his government's endorsement of the program in early 1973, Prime Minister Olof Palme noted with pride that by the turn of the century Sweden would be the world's largest per-capita consumer of nuclear power. In the shorter term, the reactor program would allow Sweden to compete for world export markets in nuclear power and would be an important source of employment and economic growth. In short, nuclear energy would increase the material well-being of this high consumption society while simultaneously decreasing its dependence on imported oil.

Prior to 1973, there had been virtually no recent public attention to the nuclear power program. In the early 1960s wide discussion of the desirability of acquiring nuclear weapons had cemented public opinion against nuclear weapons but in favor of commercial nuclear power. Shortly thereafter, government and industry began to cooperatively develop heavy water reactor systems. Their efforts culminated in the controversial Marviken plant, which was a technical failure. The ensuing scandal brought about the second major public debate about nuclear power in Sweden—not nuclear power, as such, but rather the management of the heavy water development program.

At the heart of developments between 1973 and 1975 was the Center Party. Founded in 1921 as an agrarian coalition whose objective was subsidies to farmers, the Center Party was known until 1957 as the Farmers' Party. In 1932, it reached an historic compromise with the Social Democrats that lost its voters to other political parties in all subsequent elections. Its change of name in 1957 signified a change of heart as well. Into the 1950s, its interests were tightly parochial. But with the steady decline of farmers in industrializing Sweden, the party leadership was faced with the need to broaden party appeal in order to survive. In 1957, the renamed party began to court the urban vote in earnest.[9]

The force behind the change of strategy was Gunnar

Hedlund, who became party chairman in 1949. Hedlund was widely regarded in Sweden as one of the most able and clever politicians of the postwar period. Consistently able to extract the maximum advantage from almost any situation, Hedlund kept the party flexible and alert to the changing winds of opportunity, while it pursued the interests of small, local, and decentralized interests. During the 1960s, it attempted to represent the small businessman, the small farmer, and the cause of local government. Its self-image, and the image it tried to project to Swedish voters, was that of an underdog bravely fighting the crushing impersonality of bureaucratic centralization in the interest of a more humanized society. By the late 1960s the party had become closely identified with the "Green Wave" of protest against the policies of the Social Democrats.

In 1972 the party decided that the Swedish nuclear program was an especially attractive target. Its new leader, Thorbjörn Falldin, was a farmer from northern Sweden whose voting record in the Riksdag revealed persistent suspicion of science and technology. He was reportedly greatly influenced by the highly publicized warning of Nobel laureate Hannes Alfvén that the safety of nuclear power could not be taken for granted, and by reports of the Emergency Core Cooling System controversy in the United States. Falldin took the position that nuclear power was unacceptable for Sweden, and even his sharpest critics and political opponents believed he uttered this opinion with profound sincerity.[10]

When the Prime Minister endorsed CDL's nuclear construction program, he anticipated opposition from the Center Party. The opposition turned out, however, to be extremely effective. In the Riksdag committee reviewing the government program, the Center Party representative concentrated her attention on the issue of nuclear safety. The program's sponsors were apparently notably ineffective in answering her questions, particularly those about fuel reprocessing and waste disposal.[11] On the basis of the committee's report, the Riksdag resolved in May 1973 that "no decisions in favor of further expansion of nuclear power should be taken until new basic material, including information in the state of research and development on radioactive waste disposal, has been put before the Riksdag."[12] The expression "further expansion" referred to plants over and above the 11 already authorized. Several new Riksdag commissions dealing with nuclear power were established.

The Social Democrats were not especially disturbed by these events. Like their counterparts in the political leadership of other countries with large nuclear programs, they tended to accept the assurances of nuclear industry engineers that although safety was a serious issue, reasonable people would readily concur that the potential hazards were tractable once the "facts" were made known. Throughout the summer of 1973, the Ministry of Industry continued to believe that the spring moratorium resolution would not affect its plans and those of the CDL.[13]

As it turned out, the nuclear program was not an issue in the 1973 fall national elections, but these elections did considerably change the balance of power in the Riksdag. The Social Democrats became dependent on the Communist Party *and* the support of at least one non-Socialist party. Of the 350 seats in the Riksdag, Social Democrats won 156, 19 went to the Communists, and 175 now belonged to the three non-Socialist parties: the Center Party, the Liberal Party, and the Conservative Party. Thus, the two blocs were evenly split with 175 seats each. The leadership of all the parties recognized that this was a situation in which everyone would have to be amenable to negotiations and be prepared to assume responsibility for necessary compromises.[14]

The 1973 Middle East War and the OPEC embargo closely followed the Swedish elections. As in other European countries, the electric power industry and the government leadership interpreted these events as justification for an accelerated reactor program. It soon became evident tha they had misread the strength of the opposition. In February 1974, CDL conducted a private survey of public attiudes toward energy supply and demand that shocked everyone involved: almost half of the national sample said they were "strongly committed" against any additional nuclear power in Sweden, while only 25 percent said they were willing to accept "some additions" to Swedish nuclear plant capacity. The remaining 25 percent had no opinion.[15]

Throughout the spring and early summer of 1974, media attention to the energy crisis mounted steadily, and the nuclear power program became the focus of attention. By summer, the leaders of every party had concluded that they would have to put the question to their membership. So began organized discussion by "party study groups" in accordance with the established procedures of Swedish political democracy.[16] The Liberal Party, for example, organized approximately 700 separate study groups, the Center Party some 1,000, and the Social Democrats more than 3,000. Each of these groups met several times during the early fall to discuss nuclear power and the background information supplied by the central party organizations. In all, during the fall of 1974, approximately 7,000 study groups of 10 to 20 members each debated whether they were for or against nuclear power for Sweden—extraordinarily broad and deep attention by a significant fraction of the nation's electorate. At the conclusion of each group's work, the opinions of the participants were collected by the central party organizations.

For the Social Democrats, the outcome was a nasty problem: the party members were almost evenly split, and opinions were strong on both sides.[17] Among Center Party membership, 65 percent of all participants were firmly opposed to any further nuclear power in Sweden; of the remaining 35 percent, 80 percent said no more than four or five additional nuclear plants should ever be built in the country. More than 50 percent of the Center Party participants claimed to be prepared "to make sacrifices in their standard of living if that were necessary to avoid any further commitment to nuclear energy."

Each of the five major political parties formulated a proposed energy program in light of its study-group results. The one item of consensus among all parties was that vigorous efforts would have to be made to reduce energy consumption growth rates with strong conservation programs; growth rates in energy consumption, especially electric power consumption, should be reduced by at least 50 percent. Each program also addressed the nuclear issue. The Conservative Party endorsed the government's original program "in principle" but said that important unresolved questions had been raised by the critics, particularly with respect to aspects of the light water fuel cycle. To settle these matters, a massive government/industry research and development program should be a condition for further nuclear plant construction.

The Liberal Party went somewhat further. It opposed resumption of the original program until research had settled some of the questions raised during the debate. The party said it would take no formal position on the ultimate acceptability of reactor technology pending the outcome of this further study. The Center Party flatly opposed more nuclear plants and said that serious consideration should be given to canceling those under construction. The Communists agreed, reversing their previous position and becoming the only Communist Party in the world to formally oppose nuclear power.[18]

On January 2, 1975 Prime Minister Palme announced the government's new energy policy.[19] It differed fundamentally and dramatically from the program he presented two years earlier, for it stressed energy conservation and demand control rather than modification and expansion of the nation's energy system. In his words, "the first cornerstone of Swedish energy policy must be a concerted effort to restrain consumption." Consequently, government policy would aim to reduce the growth rate of power consumption to an average of 2 percent per year between 1975 and 1985 and then aim for a constant level of power consumption by about 1990.

The Prime Minister explicitly mentioned the effects of the national debate over nuclear power:

> Only a few years ago the predominant attitude towards the peaceful uses of atomic energy was unreservedly positive. The intensive debate which has since taken place concerning the hazards of nuclear power and the problems of safety have changed this attitude. A critical attitude to nuclear power is now apparent, ranging from complete opposition . . . to attitudes characterized by doubt and uncertainty. . . . Skepticism and caution are very much the order of the day.

He summarized the outcome of the past two years of debate by noting that "a great majority accept the nuclear power program adopted by the Riksdag, at the same time calling for prudence and a reservation of judgment. *We should act on this advice*" [emphasis added].[20]

In practice, the new, cautious approach to nuclear power meant a government request to the Riksdag for authority to construct only 2 new reactors instead of the original 13, together with a pledge to reassess the role of nuclear power in Sweden in 1978. The Conservatives agreed to support this program, thereby assuring its passage over the opposition of the Center Party and the Communists on May 27, 1975.[21]

Many in Sweden, Europe, and the United States interpreted this slowdown as a considerable victory for the opponents of nuclear power. The critics themselves were not so certain. They pointed out that reactor construction would proceed at least until 1978 at exactly the pace the government proposed in 1973. Although the government was, in principle, committed to reassess the situation in 1978, there were no binding guarantees that the 13 authorized reactors were the last nuclear plants CDL would build. Accordingly, the Swedish compromise did not differ in any material way from the outcome of the French debate. In both cases, the government merely agreed to reexamine the national commitment to nuclear power two or three years later, meanwhile, proceeding as planned with reactor construction.

This negative interpretation is superficial. After all, many officials in Swedish government and industry who believed nuclear energy to be a panacea for the national energy problem were surprised and impressed by the public's response to the arguments of the nuclear critics. Some of the strongest supporters of nuclear power readily conceded that it had been embarrassingly easy for the Center Party to raise serious questions in the minds of intelligent people, including a surprising number of members of Parliament from other parties. They also conceded that the government badly misjudged the May 1973 moratorium resolution, which it should have taken as a signal that trouble lay ahead. This resolution expressed the collective judgment of the Riksdag that nuclear advocates had left legitimate questions unanswered.

Taking the Issue to the People: The 1976 Referenda in the United States

It had long been an article of high faith within the American nuclear industry that as soon as reasonable people were given an opportunity to examine all the pertinent evidence, their doubts about the benefits of nuclear power would vanish. Apparently, in the nuclear debate in Sweden, just the opposite happened.

In California, during the summer of 1975, opponents of nuclear power gathered enough signatures to initiate a statewide voter referendum which, if approved, would have stopped construction of new nuclear plants in the state. The vote was to take place during the state's presidential primary election in June 1976. The nuclear power question became known, due to its position on the ballot, as "Proposition 15."

As 1976 began, it appeared that the outcome of the plebiscite would be very close. A number of public opinion polls showed California voters to be one-third in favor, one-third against, and one-third undecided on Proposition 15. The nuclear industry's faith in the persuasive power of its case would be put to a severe test, and the events in Sweden and elsewhere in Western Europe hardly promised the industry a happy outcome.

There were other reasons to suppose that the supporters of Proposition 15 were in an extremely strong position. American political parties have no mechanism even remotely resembling the Swedith discussion groups for mobilizing the opinions of their members. Hence, the nuclear industry could not look to the major parties—both of which supported the industry's position—for much help in opposing the referendum. This inability of American parties to deal with specialized issues such as Proposition 15 has often allowed highly motivated minorities operating outside the formal party structures to control election outcomes on specific matters they either favored or opposed. To achieve their objective the nuclear opposition in California merely needed to split the "undecided" voters more or less evenly with industry. The Swedish debate suggested that accomplishing this goal would not be too difficult. Decisive numbers of previously uncommitted citizens had been receptive to the Center Party's claim that the nuclear industry bore the responsibility for resolving outstanding nuclear safety questions.

Proposition 15 was put to the test on June 8, 1976, and the nuclear opponents lost by a two-to-one margin. Evidently almost all of the undecided voters rejected their appeal. Five months later, similar referenda were held in several other American states during the presidential election. In every one of them, with remarkably little variation, the referenda were defeated by about two-to-one. Why? Advocates of nuclear power would like to interpret these results as proof that two-thirds of American voters find nuclear power acceptable. This is, of course, plausible. But our own view differs.

Despite the apparent similarity of the issues debated in America and Sweden, the advocates and opponents of nuclear power in the United States presented very different cases to the voters than had the Swedish parties.

One of the central themes of the public relations campaign of the nuclear industry and electric utilities was the link between nuclear power and economic prosperity. In well-financed advertising campaigns, they characterized the attack on nuclear power as an attack on a program vital to a growing, full-employment economy. They pictured critics of nuclear power as misguided or irresponsible intellectuals more concerned with remote chances of accidents than with the clear and present dangers of economic stagnation, high unemployment, and a dangerous reliance on foreign oil.

For their part, the nuclear critics largely failed to effectively insist that it was industry's responsibility to answer fully specific questions about the hazards of nuclear power. They did not insist, for example, that the industry present detailed plans and timetables for nuclear waste disposal or for increasing the security of fissile material. The nuclear opposition apparently forgot that it had to persuade uncommitted voters to give it the benefit of the doubt on specific technical issues. Instead, by attacking "big business" and "corporate monopolists," the critics merely made more credible the industry's claim that they really were attacking the institutions which had provided growing incomes to middle-class wage earners.

In our opinion, however, the task of nuclear critics in America is far more difficult than in a society like Sweden, which has a long history of left-wing opposition to establishment policy. When the nuclear critics in California and other states entered the political arena, they were obliged to contend with a far less congenial environment than they had found in the administrative/judicial apparatus. Even after

Vietnam and Watergate, the traditional symbols of authority in American society retain great legitimacy for most citizens, certainly for more than a majority of the electorate. By appearing to attack these symbols on an especially sensitive issue—jobs and income—the nuclear opposition isolated itself from the mainstream attitudes whose support was crucial to success with the voters.

In the aftermath of these referenda most national political leaders believed that the burden to be specific was still on nuclear critics. The establishment still appeared to accept the industry's claim that all the problems identified either had been or could be resolved. In areas of admitted deficiency —long-term waste disposal and safeguards against the theft of plutonium—there was a similar willingness to accept the industry's assumption that every problem has at least one solution. In fact, the industry's most important asset may be the range and depth of American agreement, explicit or implicit, with this proposition. This faith may also be one of the industry's most basic points of disagreement with its critics.

The reactor safety controversy in the United States was by no means resolved by the 1976 referenda. The nuclear opponents had clearly raised doubts in the minds of a large, though not electorally decisive, number of observers with no prior stake in the issue. Contrary to many expectations, a widespread debate and an increase in available information appeared to have enhanced doubts and uncertainties about nuclear power for a large fraction of the American public. Unconditional support for the industry position seemed increasingly concentrated among either the previously committed or the relatively unimportant.

Beyond the Technical Argument

A characteristic of the nuclear power debate which especially stands out in the Swedish experience is that controversy is hardly confined to facts. In Sweden, both sides recognized that they had more on their hands than a technical disagreement over safety when technical arguments on reactor safety were met with comments on the quality of life.

Contrary views about the relation between technology and society emerged. First was the position that material progress cannot be postponed. Said one Swedish student of the controversy:

> We who remember how things were do not want to go backward. Only those who have no experience can dream of that. The prospect of further material progress is in clear and present danger due to the insecurity of national oil supplies. This situation must be rectified. In practice we must look for solutions which are available today. Nuclear power plants are a necessary part of such a solution.[22]

On the opposite side were two schools of thought which stressed more than material progress. Some argued that new technologies can help maintain material progress but that they must not be allowed to dominate other social purposes, especially environmental protection. The conflict between the two values is particularly acute in the case of nuclear power. Until nuclear power is demonstrated to be environmentally acceptable, Sweden must choose to stop material progress.

Others took the more extreme position that it is not necessary to further enhance material standards. Quite the contrary, they saw such "progress" as the cause of many of Sweden's present problems: pollution, increasing crime rates, and so on. The important, clear, and present danger to Swedes is the threat to their human dignity by a society too complex.

Nuclear power technology is a manifestation of inhuman values carried to the point of insanity.

These arguments underline a fundamental truth about the debate over nuclear power in the Western democracies. Although it is far more evident in Europe, it is nonetheless also true in the United States that the debate is, in part, really a discussion about the nature of contemporary society.

Part III

THE TWILIGHT OF PROBABILITY

CHAPTER 9

The Economic Swamp

THE TRANSNATIONAL nuclear safety imbroglio is directly connected with the economic developments which we reviewed in Chapters 4 and 5. There we saw a record of continuing increases in the cost of electricity from light water reactors. Some 20 years after the first light water reactor began to produce electricity, the cost had yet to stabilize. Some 10 years after the beginning of debate about the relative economic merits of coal and nuclear power plants, the issue had become more, not less, obscure and complex.

In the early years of light water reactor commercialization, it was plausible to attribute cost increases to the initial optimism which invariably accompanies a major technological innovation. Perhaps there was no sound reason to suppose that the disappointing early "economic returns" from nuclear power were very much different from similar initial disappointments with other advanced technology products.

However, it was clear by the end of the 1960s that something special was happening in the reactor commercialization process: the establishment and hardening of a circular flow of misinformation on the economic status of light water reactors.

This merry-go-round perpetuated the belief that the causes of prior cost increases were well understood and had already been brought under control.

By the early 1970s, two rather different explanations had developed in the United States about what had gone wrong with the economics of nuclear power. One was common among the purchasers of light water reactors; the other was generally held by the government officials who were promoting and regulating the utilities. The electric utility industry generally placed the blame on the government's environmental protection policies, quality assurance requirements, and nuclear safety regulations. The government countered by pointing to the industry's poor labor productivity, manufacturing failures by the equipment suppliers, and management problems in plant construction.[1]

To support their argument, the electric utilities were able to cite an apparently endless series of changing regulatory requirements that had begun with the first commercial orders for light water reactors. For example, in 1971 the Atomic Energy Commission issued radically new radiation protection criteria which required that nuclear plants be designed to hold all radioactive emissions during normal operation to levels which were "as low as practicable."

To support its position, the government could cite a steady increase in the number of worker-hours per kilowatt required to build a nuclear power plant. Plants which entered operation in the early 1970s generally required three to four worker-hours per kilowatt to complete; those then under construction would require more than double this figure.[2]

These differences of opinion were the subject of numerous discussions within the nuclear power community during the early 1970s. To industry representatives, the solution to the problem seemed straightforward: the government merely needed to "freeze" its regulatory requirements. Then "standardized" nuclear power plants could be designed and sold; "learning" would inevitably occur; and costs, finally, would start to decline. Government officials did not question the desirability of standardized designs; indeed, they argued that industry should get on with the job of producing them. But they insisted that there was also a need for improved management of plant construction and for more rigorous quality control by both sellers and buyers of light water reactors.

The unfortunate thing about these discussions is that they minimized the real causes of the problems which they purportedly addressed. After more than a decade of experience with large light water nuclear power plants, important engineering and design changes were still being made. This is contrary to experience with most other complex industrial products. The failure of light water technology to be standardized is believed by many in the industry to be the result of extensive perfectionism and conservatism on the part of both customers and government regulators. But in our opinion there is another, more fundamental, reason for this relatively unusual situation.

For 15 years many of those most closely identified with reactor commercialization have stubbornly refused to face up to the sheer technical complexity of the job that remained even *after* the first prototype nuclear power plants had been built in the mid- and late 1950s. Both industry and government refused to recognize that construction and successful operation of these prototypes—though it represented a very considerable technical achievement—was *the beginning and not the near completion* of a demanding undertaking. During the late 1960s and early 1970s it became apparent to many in the nuclear power industry that another key ingredient was critical to successful innovation: operating experience. Only actual experience with full-scale plants and facilities could provide the basis for design modifications needed to guarantee plant reliability and safety. In the important years following the Great Bandwagon Market, it became painfully evident that the problems associated with building and operating 1,000 to 1,200 MW nuclear plants bore disappointingly slight resemblance to those associated with 100 to 200 MW plants.[3] The cost of this lesson has been high.

It is also now apparent that the "learning by doing" process is particularly slow in the nuclear power business, and that the benefits of it are not easily transferred from one power plant to the next,[4] much less from one country to another. Many European reactor manufacturers recognized this only quite belatedly, and again at high cost. In retrospect, it was naive of Europeans to believe that by rushing ahead to get on the light water bandwagon in the late 1960s they would benefit from American experience and avoid "re-learning." By the mid-1970s it was clear to both reactor manufacturers and electric utilities in Europe that foreign construction experience was only marginally relevant to their own needs.[5]

The blunt fact is that 15 years after the Oyster Creek sale, light water technology has still not attained the technical maturity which its promoters thought it had reached years earlier. But, in addition, to be fair to the nuclear industry, external circumstances were partly responsible for the failure of the technology to mature in spite of considerable construction and operating experience. There is a continuing disagreement over "how safe is safe enough?" In fact, by the mid-1970s it was clearly this dispute which was at the center of the economic woes of nuclear power.

An example will illustrate the truth of this observation. In the early years of light water plant construction, the structural steel and concrete required per kilowatt of generating capacity decreased sharply and steadily. In 1955, a 180 MW light water reactor design called for more than 30 tons of structural steel and about one-third of a cubic yard of concrete per megawatt. By 1965, a much larger plant of about 550 MW required less than half as much of these materials per megawatt of capacity. These efficiencies reflect classic "economies of scale." Then, in the late 1960s, the trend reversed. Larger light water plants began to require more, not less, structural material per unit of capacity; by 1975, the steel and concrete needed per megawatt for a 1,200 MW plant approximately equaled the 1960 requirement for a 200 to 300 MW design.[6] This reversal was a direct consequence of stricter safety and environmental protection requirements laid down during this period. More stringent safety requirements meant thicker concrete walls.

Aside from the direct material costs of additional safeguards, the nuclear critics have affected nuclear power plant costs in three principal ways. First and most obvious, they have slowed down the plant licensing and construction processes. The time required to build a light water plant in the United States has consistently grown since 1965. During the Great Bandwagon Market, 60 to 70 months were required between the time a reactor was ordered and when it was connected to the electricity network. By 1975, this period had approximately doubled.[7] It is more costly to build a plant over 10 years than to build the same plant in six years. Delays invite unanticipated inflationary effects because costs are actually incurred later than originally anticipated. They also bring about additional interest on the funds utilities borrow to finance construction. Finally, direct construction costs are also affected by delays. In 1973, a New England electric utility estimated that a year's loss in schedule for an 800 MW nuclear plant approaching completion would cost $30 million in extra interest payments and $10 million in extra overhead charges.[8]

The second contribution of the critics to cost escalation is in their insistence that partially and wholly completed plants be modified to meet the safeguards they have imposed. In late 1972, the Atomic Energy Commission's Advisory Committee on Reactor Safety received an anonymous letter raising questions about the location of certain piping in the Prairie Island plant in Minnesota. After review of the information submitted, the Commission discovered that in the event of a rupture in certain locations the routing of some portions of the main steam line threatened damage to nearby essential safety equipment. On this basis, the Commission determined that changes must be made at the Prairie Island plant. It later required all other utilities with nuclear plants under construction or in operation to assess the effect of such postulated pipe ruptures on their own facilities. At the time of receipt of the letter, Northeast Utility's Millstone #2 plant was 50 percent complete. To meet the new government requirements, all stress analysis calculations for the piping system were recalculated and substantial design modifications made after most of the piping systems and electrical equipment had already been installed. The utility estimated this one change to cost about $5 million.[9]

A third reason why nuclear critics are responsible for the cost increases is that they have sponsored the burgeoning requirements for environmental assessments and reports of various kinds, together with the proliferation of state and local siting boards and panels. The increasing administrative and legal complexity of plant licensing adds significantly to costs. Prior to 1972, government regulations allowed applicants for nuclear plant construction permits to conduct site exploration and excavation activities prior to receipt of the permit. Moreover, before that time, the Atomic Energy Commission liberally granted exemptions to certain related rules. In practice, this policy helped utilities accomplish one to two years of construction prior to receipt of a construction permit. Under more recent government rules, this head start has became illegal. This single change in federal regulatory practice has, therefore, lengthened the time between issuance of a construction permit and fuel loading by many months.

In their discussions of nuclear power economics during the early 1970s, industry and government misinterpreted the significance of the safety controversy. The nuclear power community's faith that opposition was a transitory phenomenon clouded its diagnosis of the problem and led it to focus on symptoms. Complaints about the failure of the regulatory process were made as if these failures were little more than bureaucratic capriciousness or poor management.[10] Likewise, the government berated the industry to get on with the job of standardizing its product, as if its past failure to do so represented a semirational preference for change over stability. The entire protracted exchange was remarkable chiefly for its apparent obliviousness to the existence of the *growing*, not diminishing, nuclear opposition whose objectives were directly opposed to those of both government and industry. The nuclear critics' objective was to complicate, not to simplify, the regulatory process; it was to enhance, not to diminish, the pressure on government to continue changing regulatory requirements; it was to slow down, not to speed up, the licensing and construction of nuclear power plants.

At the beginning of the reactor safety controversy, critics of nuclear power simply took upon themselves a responsibility which, with some justification, they believed the American government had abdicated. Individuals with a variety of objectives began to ask questions about the safety of the plants then being sold on the open market. As a consequence the plants currently being built are safer than those designed in the mid-1960s. In the 1970s it has remained the advertised purpose of most of the American critics merely to guarantee society an acceptable reactor technology. With the ostensible intent of making nuclear power "more acceptable," they have for more than a decade imposed on industry and government their own often rather imprecise views about "how safe is safe enough."

The nuclear establishment's discomfort with outside criticism was, therefore, partly justified. It is far easier for an outsider to *claim* that there is a possible risk than it is for the industry to *prove* that there is none. The nuclear industry was also surely correct in its belief that nothing would persuade certain members of their opposition that nuclear power is acceptable. Some nuclear critics in the United States and abroad are plainly determined to oppose nuclear power until the technology is abandoned, regardless of any pertinent technical or economic developments. Perhaps it is understandable, then, that the nuclear industry found it hard to imagine that the cost of nuclear electricity might never stabilize, let alone decline, as long as critics were able to harass the industry in the regulatory process.

By the mid-1970s, the critics actually began to point to the economic problems they had brought about as new cause to junk nuclear power reactors. They pointed to the apparent failure of light water reactors to establish decisive economic superiority over coal in spite of a quadrupling of fossil fuel prices. At the instigation of nuclear critics, the economics of nuclear power became an additional issue in licensing proceedings. The result has been a series of self-fulfilling prophecies. Critics of nuclear power have demanded a review of generating economics during licensing proceedings. Of course, the time-consuming process of a thorough review contributes to a negative report.

By the end of 1976 the same apparent technical confusion which had long characterized the reactor safety issue had also overtaken nuclear power economics: it was virtually impossible to make any positive claim without offending one of the parties in the dispute. The critics had developed their argument along a number of different dimensions. Nuclear advocates, of course, had well-rehearsed counter-arguments.

The controversy touched every important factor on which the economic performance of light water plants depended: the relative investment costs of coal or nuclear generating stations; the factors involved in allocating this cost to each kilowatt hour of electricity produced by the plants during their assumed lifetime; and even the appropriate way to handle inflation in figuring generating costs.

The Relative Investment Costs of Coal and Nuclear Plants: An Endless Debate

Some two years after the OPEC embargo, most informed persons acknowledged that the cost of building light water power plants had still not stabilized. But the nuclear industry increasingly insisted that coal plant costs were now climbing at equivalent rates. There was room for a great deal of honest disagreement because this is a genuinely complex issue and there was surprisingly little pertinent data. In mid-1974 the rather scanty evidence that was available suggested that during the prior five years, the costs—in constant dollars—of nuclear power plants had grown at more than twice the rate of coal plants. Nuclear critics elaborated, claiming that the competitive position of nuclear power had been deteriorating and that the fuel cost advantage of nuclear power was already offset by a widening capital cost disadvantage.

The nuclear industry did not seriously dispute that the evidence of the recent past was discouraging. But the really contentious question was what would happen in the late 1970s and the 1980s. The nuclear advocates insisted that adverse economic developments were inevitable for coal. By the mid-1970s, American electric utilities were all subject to strict federal and state air quality control regulations.[11] Everyone agreed that meeting these standards would sharply increase the costs of building and operating coal-fired power plants. Just how much it would adversely affect these costs was another matter. The available data allowed a broad band of equally plausible guesses. Naturally, the nuclear industry usually chose the most pessimistic—for coal—while the critics preferred the most optimistic.

Nuclear Plant Performance: More Arguments from Inconclusive Evidence

The total number of kilowatt hours of electricity that a generating station can produce over its lifetime is crucial to its economic performance: the more kilowatt hours, the lower the fraction of the capital cost that must be allocated to each. Power plant performance is measured by a number known as the "capacity factor." It is simply the ratio of the kilowatts produced during a given period to the total which a plant operated at constant full power could theoretically produce during the same period. The first commercial light water reactors were sold with the promise that they would achieve capacity factors of 80 percent or higher. This was soon recognized to have been highly optimistic. Though early technical difficulties with large light water reactors considerably diminished those expectations among both reactor manufacturers and utilities, the capacity factors which have been effectively realized by light water reactors have developed into another acrimonious issue in the nuclear power controversy. Nuclear critics have contended that, on the average, light water reactors in the United States have operated at less than 50 percent capacity. Alternatively, it is the general position of the manufacturers in the utility industry that realized capacity factors have been in the neighborhood of 65 percent.[12]

This might seem to be a relatively simple matter to resolve, for there is no conceptual difficulty with the notion of a capacity factor and it would seem to be a fairly easy thing to measure. However, the wide range of variation in the performance of different plants in operation and the relatively short time from which information is available about the newest and largest plants has produced a situation which is more suited to rhetoric than to analysis: the available data are so thin and the range of variation so broad that widely differing judgments can be supported in seemingly reasonable ways.

The Result: Continuing Economic Indeterminancy

Three years after OPEC appeared to have saved the light water reactor industry, it was still possible to argue that these machines were either "competitive" with coal-fired generating plants, "almost competitive," or "not yet competitive." Moreover the range of apparently reasonable relative cost estimates had become wider than ever before. Consider first the capital cost of a nuclear power plant. Authoritative sources differed by about 100 percent, from low estimates of about $400/kw to high estimates of about $800/kw (constant 1976 dollars).[13] Estimates of the ratio of coal-fired plant capital costs to those of nuclear plants also varied widely, principally because of a continuing lack of reliable cost data on equipment to control sulfur emissions. Because the available information was still so poor, it was possible to maintain that new coal plants might either cost essentially the same as nuclear plants on a per kilowatt basis, or if relaxation of government air quality control policy was assumed, that they would be as much as one-third cheaper.[14]

The "fixed charge" rate used by electric utility companies to depreciate their capital investments was also open to debate. Numbers ranging from 15 percent to 20 percent were proposed.[15]

The especially contentious argument over capacity factors has already been noted.

Taken together, what all of these disagreements meant was that of the four numerical terms needed to calculate simply the capital-related component of the relative cost of electricity from nuclear or coal plants, only one was not in dispute: the number of hours in a year (8,760).

It is important to have some quantitative sense of just how large the resulting band of economic indeterminancy had become. To do so, let us pick two sets of assumptions from the available menu. We can call the first of these sets "pessimistic" for nuclear power. They might be:

- Nuclear power plant capital cost: $750/kw (constant 1977 dollars)
- Coal-fired plant capital cost: 0.70 × nuclear = $525/kw
- Nuclear plant capacity factor: 0.60
- Coal-fired plant capacity factor: 0.67

An alternative set of assumptions which we can label "optimistic" for nuclear power might be:

- Nuclear power plant capital cost: $550/kw (constant 1977 dollars)
- Coal-fired plant capital cost: 0.85 × nuclear = $467/kw
- Nuclear plant capacity factor: 0.80
- Coal-fired plant capacity factor: 0.60

These capital-related terms do not, of course, exhaust the list of items which are uncertain in the nuclear vs. coal cost equation. We have already mentioned the range of opinion about the appropriate value of other capital-related term, the "fixed charge" rate. But we have not said anything about the fuel-related components of the cost of electricity.

Historically, the so-called "fuel cycle" cost of nuclear power was the one apparently stable element in the economic situation.[16] That, too, had changed by 1977. The price of uranium sky-rocketed in the early 1970s and future developments depended, among other things, on the outcome of multi-hundred million dollar law suits. The nuclear non-proliferation policy of the Carter Administration in the United States had introduced at least proportional uncertainty into the estimation of the future costs of spent nuclear fuel management, fuel reprocessing (if allowed at all), and nuclear waste isolation.[17]

But, let us return to our two sets of capital-related cost assumptions. If we work with the "optimistic" set, apply a low fixed charge rate of 15 percent, a coal plant heat rate of 10,000 btu/kwh, *and* assume further that nuclear fuel-related costs remain close to their historic levels, then nuclear power can be argued to be competitive with coal which is as cheap as $.44/million btu.

Alternatively, if we work with the "pessimistic" assumptions, choose a fixed charge rate of 19 percent, a coal plant heat rate of 9,000 btu/kwh, and suppose that the fuel-related costs of nuclear power double, then nuclear power cannot compete with coal which is as expensive as $2.46/million btu. (This figure climbs to $3.12/million btu if nuclear fuel costs triple—a development which is unlikely, but not wholly improbable.)

None of these numbers should be taken literally. Their purpose is to illustrate the fact that it is literally impossible to make a determinate economic case either for or against nuclear power. For naturally much of the real debate about the economics of nuclear power is not between extremely "optimistic" and extremely "pessimistic" relative cost estimates. It is based, instead, on fairly small differences in assumptions between the two extremes. For instance, is the capital cost of a nuclear power plant $600/kw or $650/kw? Is the appropriate capacity factor 0.60 or 0.65? And so forth. Hence the argument is usually about differences in the estimated cost of electricity from coal-fired or nuclear plants of "only" a few mills per kilowatt-hour—a few tenths of a penny. But, because a large power plant produces billions of kilowatt-hours per year and is expected to operate for thirty years, the debate is really about the possible "savings" or "loss" of many millions of dollars, even many tens of millions of dollars for a single plant. Understandably, this kind of issue seems to be important to many persons—private citizens as well as business executives and public officials. This shared sense of importance has produced a bewildering, frustrating, and apparently endless public argument between those who hold an optimistic view of the relative economic prospects of nuclear power and those who hold a relatively pessimistic view. The argument often goes on before regulatory commissions and even in the courts: an adversary environment which is particularly ill-suited to the productive discussion of an essentially indeterminant question.[18]

The elusive quest for an answer to the economic case for nuclear power has led us into a swamp where two warring parties wrestle to a stalemate with widely differing assumptions and inconclusive facts.

CHAPTER 10

The Political Stalemate

IN THE WORLD of OPEC energy economics, the price of energy has little to do with the cost of extracting it from natural resources. Instead, the price of energy is now determined by relative values assigned to energy-producing resources and technologies by those who either control or require them. The price-setting process for energy has become highly political. Accordingly, the concepts of political analysis—constituencies, legitimacy, and power—are as important in understanding what is going on as the elements of economic analysis: markets, elasticities, and costs.

Most of the world's governments have accepted the need to develop energy policies. Their purpose is commonly formulated in terms of bringing demand and supply into balance, subject to various technical, economic, or political constraints. But in the public debate over these policies, political constraints are often assumed to be somehow less real or less worthy of consideration than others: business and government leaders in the United States and Europe have believed that political opposition to nuclear power would vanish as soon as the public recognized that it was the cheapest way to meet a large part of future energy requirements.

This formulation reverses the real causal sequence. Critics

of nuclear power have, in effect, contended that its apparently cheap electricity disguises its true social cost. They state that the true value of power reactor technology has been overstated because the costs of various hazards have not been considered. In fact, some nuclear critics go so far as to claim that reactors can never have a positive value, that nuclear fission can never be an acceptable way to produce energy no matter how high the price of alternative sources. In the United States, those who hold these views have successfully used the administrative and judicial machinery of American democracy to bring about delays and cost increases too great for the reactor manufacturers' customers to ignore.

The result has been a sharp decline in the number of nuclear power plants ordered in the United States since 1973. A similar situation has developed in most other Western countries. There has been an apparent worldwide halt in new orders for nuclear power plants. During 1976, worldwide reactor sales included only eleven new units: three in the United States; four in France; and two each in Sweden and South Africa. In all Western countries, a significant increase in the activities of the nuclear opposition coincided with the 1974–1975 economic recession. For the American electric utility industry, the leveling off of demand during the recession provided welcome respite. Planning the construction of a nuclear power plant had become a costly and burdensome undertaking at best, a highly uncertain one at worst. It was likely to mean a pitched battle with an entrenched and dedicated opposition that would drain away financial and managerial resources.

These events meant that President Carter's new administration inherited a stalemate on nuclear power. The nuclear opposition had been unable to win a decisive electoral or legislative victory, but industry and government had been unable to affect the opposition's ability to use local, state, and federal licensing procedures to cause delays and cost overruns. By early 1977, it was evident that the result of this stalemate was a *de facto* moratorium on the purchase of new nuclear electric generating equipment.

Some three years after the political leaders of most oil-importing countries tried to make nuclear power the cornerstone of new energy policies, it seemed unlikely that any democratic society could avoid the articulation and growing appeal of antinuclear views. The political power of those who held these opinions could not be ignored by the governments of any of the Western democracies. Public opinion surveys in the United States and Europe revealed a generally consistent pattern: between one-quarter and one-third of those questioned expressed some reservations about the safety of nuclear power; an approximately equal fraction had no such reservations; and the rest had no discernable opinion at all.[1]

The means used by the antinuclear movement to achieve their objectives have varied widely generally depending on the opportunities available for intervention in the plant licensing process. With the exception of Germany, no other country has provided the critics of nuclear power with the legal advantages they have in the United States. This is especially true in France, where there is literally no equivalent to the public participation and intervention that we find in the United States.

In France, however, the very absence of such opportunities, combined with the government's determination to limit outside interference with its nuclear power plant construction program has produced what appears to be a very explosive situation. The dramatic events at Creys-Malville in the summer of 1977 suggest that the government's success in limiting interference was purchased at a high political price. Louis Puiseux, a senior planning official at Electricité de France, comments on the situation as follows:

> This "hot summer" will be a watershed in the history of the antinuclear movement. It reveals France as the place where the maximum social contradictions have been produced by uncontrolled technological development.[2]

There is one feature common to the international nuclear safety controversy. This is the demand of the nuclear critics for power plant safety standards which are more stringent than those believed necessary by most persons in the nuclear industry. One of the principal effects of the American nuclear opposition has been the revision of numerous safety standards for light water reactors. It is highly unlikely that any Western nation will be politically able to build and operate light water reactors under safety standards different from those required in the United States, however "reasonable" or "unreasonable" the cost-effectiveness of such standards seems to the government concerned. For this reason, increases in the cost of electricity from light water reactors seem probable in all countries which have chosen to develop this technology.

This reaction is already evident in France. In spite of the strong organizational advantages flowing from Electricité de France's control of the entire reactor construction process, the relative economic advantage of nuclear over fossil fuels has been sharply declining in France since 1975 (Figure 10-1). Similar trends have become apparent in other countries.

The growing pressure from nuclear critics for a reduction in the level of various nuclear "risks," combined with more realistic assumptions on light water plant costs and performance, has produced a steady escalation of nuclear power costs.

The market dominance of light water reactors has also caused the energy supplies of much of the rest of the world to be inevitably linked to America's reactor safety imbroglio, and hence to American politics. To many Europeans it seems especially unfortunate and ironic that certain technical characteristics of light water reactors make them an easy target for antinuclear critics. In retrospect, it seems to them that greater reliance on their heavy water or gas-graphite systems would have reduced their vulnerability to politics in the United States.

But even if they had done what this hindsight now suggests,

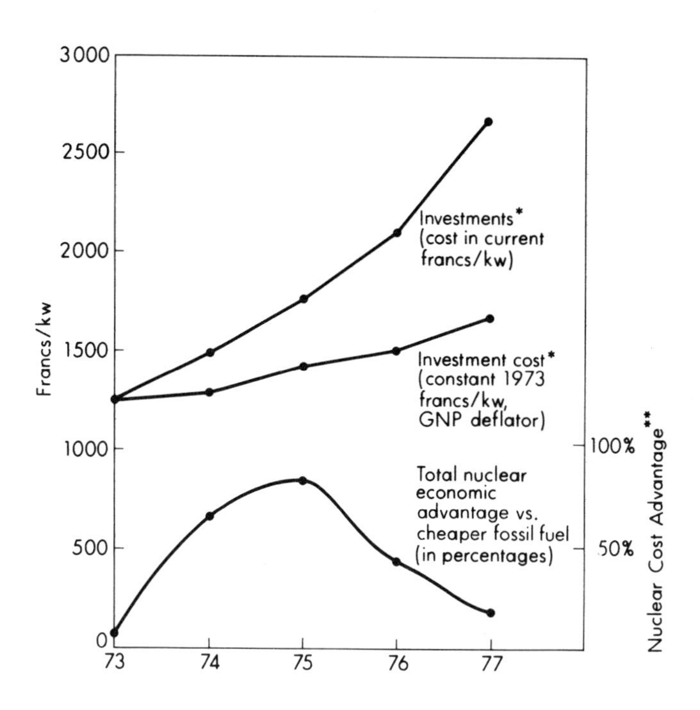

*Source: Electricité de France, and various other official sources.

**$\frac{\text{Fossil cost of kwh} - \text{Nuclear cost of kwh}}{\text{Nuclear cost of kwh}}$

Figure 10-1

the Europeans would still find themselves vulnerable to two more immediate problems: their energy poverty and the safety debate. On these grounds, the energy situation in Western Europe and Japan for the next few years is likely to be characterized by increasingly extreme positions on the usefulness or the futility of nuclear technology. On the one hand, most government officials will probably retain their strong belief in the need to develop nuclear power, but their critics will continue to paint this policy as unacceptable, undemocratic, and unprecedentedly dangerous. These are the conditions for a stalemate.

Carter's Initiative: Deepening the Stalemate?

In April 1977, the Carter Administration announced the outlines of its new energy policy. For nuclear power, the policy was an evident endorsement of a Ford Foundation-sponsored report, which had been published a month earlier: *Nuclear Power: Issues and Choices.*[3] In line with the conclusions and recommendations of this study, Carter announced that the United States would "indefinitely defer" the reprocessing of irradiated light water fuel and the use of plutonium as a supplementary fuel. Government research and development on plutonium-fueled breeder reactors would also be slowed.[4]

One of the report's authors, Harvard Professor Joseph Nye, accepted a State Department position which carried wide-ranging responsibility for policy matters connected with nuclear weapons proliferation. Another author, Harold Brown, had already been named Secretary of Defense. Their report had been the most academically respectable statement so far that the immediate use of plutonium as a reactor fuel would needlessly enhance the risk of nuclear weapons proliferation. Even prior to the publication of the Ford study, a large and technically sophisticated body of opinion had crystallized in support of this argument and it had already affected government policy. In October 1976, President Ford announced that "reprocessing and recycling of plutonium should not proceed unless there is sound reason to conclude that the world community can effectively overcome the associated risks of proliferation."[5]

The authors of the Ford Foundation study offered several arguments in support of their recommendations. Most novel was their contention that reprocessing and plutonium recycling could be deferred and the introduction of the breeder delayed without causing any significant economic penalty to nuclear power for years, and perhaps for decades. They claimed that any economic benefit from these innovations during this century was "questionable." This contradicted more than 10 years of conventional thought about the place of plutonium in the worldwide development of nuclear power.

Plutonium was to be an important part of the world's post-OPEC energy supply system. One of the main objections of nuclear scientists and engineers to light water reactors has always been their profligate use of uranium. But this can be partially corrected by supplementing their standard enriched uranium fuel charge with plutonium. More important, plutonium would be the basic fuel for a new generation of advanced reactor types—the "breeder" reactor—which would soon complement light water reactors in electricity generating networks.

The technical advantages of introducing the "plutonium economy" were also very straightforward. Total demand for natural uranium can be reduced by about 20 percent by "recycling" the uranium recovered from the irradiated fuel bundles. The use of the by-product plutonium as a supplemental fuel can result in additional natural uranium conservation of perhaps 15 percent. Alternatively, the spent fuel from about 20 light water reactors operating for one year would contain enough by-product plutonium to charge a breeder reactor. Moreover, the recycling of plutonium could reduce by perhaps 15 percent to 20 percent the demand for the "separative work" of enriching uranium. Total capital costs on expensive enrichment facilities would be reduced. The reprocessing of light water reactor fuel is a necessary precondition to realization of any of these economies.

In contradiction with these conventional beliefs, the Ford study concluded that the time when breeder reactors might be economically competitive was "sufficiently distant" that "the recovery of plutonium is not economically justified for many years." In any event, spent light water reactor fuel could be stored so that plutonium could be recovered later "if breeder reactors are actually deployed in the future."[6]

The study also maintained that deferring reprocessing and plutonium recycling would make the waste management problem less complex. During reprocessing, these wastes would be converted from "relatively easy-to-manage spent fuel" into a number of new forms: "high-level waste, acidic liquid waste, cladding hulls, process trash contaminated by plutonium, and others." Experience had already shown that all of these new forms create opportunities for "waste management failures."[7]

> While it has been commonly believed, particularly abroad, that reprocessing to remove plutonium decreases long-term hazards of waste, we have concluded that any reduction in long-term risk is small in comparison with the more immediate risks potentially arising in reprocessing and in the use of plutonium in the fuel cycle.[8]

All in all, they concluded, the "international and social costs" of reprocessing and recycling—of the plutonium economy—"far outweigh economic benefits, which are very small even under optimistic assumptions."

Because it questioned what everyone had believed to be two elements that were the key to assuring the long-term contribution of atomic energy, the Ford-Mitre study was widely interpreted as a fundamental challenge to the place of nuclear technology in society.

We see it, however, as a basic reaffirmation of "Atoms for Peace." The old slogans and expectations have been modified in only one important way. The Carter Administration has explicitly attempted to come to terms with the inherent linkage between electricity from nuclear reactors and atomic weapons from plutonium. It has recognized that the worldwide nuclear power program endorsed by its predecessors, which it too regards as both necessary and desirable, is an acute threat to the 30-year American effort to control the spread of atomic weapons and hence, to the U.S. conception of the international military and political equilibrium.

This recognition has created a troublesome dilemma: how to retain nuclear power as an answer, first to OPEC and, in the longer term, to scarce petroleum, while simultaneously keeping atomic bombs out of undesirable hands. The answer of the American administration was to defer light water fuel reprocessing and introduction of plutonium-fueled breeder reactors. This, it is argued, will "buy time" to establish suitable international or multinational control mechanisms and, perhaps, to reengineer the industrial chemistry of light water reactor fuel reprocessing to be more "proliferation resistant."

The Carter Administration concedes two potential flaws in its policy. The first is that its long-range (i.e., more than 10–15 year) success is highly sensitive to the world-wide supply of natural uranium. The policy requires the availability of modestly priced uranium ore for at least the rest of the century in amounts which expert opinion has hitherto considered at the outer limits of, or even beyond, what is realistic. But everyone admits that current knowledge of world-wide uranium ore suppliers is very thin.

The second potential flaw which its supporters concede to the Carter policy is that government and business interests in other oil-importing countries—and particularly in France and Germany—may be unwilling or unable to accept the potential economic or national security penalties threatened by deferring spent fuel reprocessing and commercial introduction of breeder reactors. There are substantial grounds for these concerns.

First, basic differences exist in the availability of fossil fuels between most Western European countries and the United States. The economic consequences of OPEC's quadrupling of oil prices were far more acute for Western Europe and Japan than they were for the United States. It can even be argued that these price increases strengthened America's economic advantage over most other Western countries. For most of Western Europe and Japan, the nuclear option represented a means of partially correcting this development. Nuclear power would allow these countries to regain some energy autonomy and help their balance-of-payments problems. The Carter Administration's policy of restraining the onset of the "plutonium economy" directly challenged these aspirations. The governments of Western Europe and Japan are correct in stating, "Because of its fossil fuel resources, the United States can afford to delay or even forego the nuclear option. We cannot."

Second, consider the industrial plants Western Europe and Japan developed in pursuit of the nuclear option. A general absence of indigenous uranium supplies enhanced the incentive for most Western European countries to press the commercial development of irradiated fuel reprocessing, plutonium recycling, and plutonium breeder reactors. They made substantial technical progress compared to American results. It is easy to see how the new American policy was interpreted as a self-serving attack on European efforts to come to terms with OPEC. It is also easy to understand why the Carter policy was interpreted as an attempt to prevent further development in an area of European technical and, therefore, potential commercial superiority.

Last, Carter Administration policy has had an especially damaging political effect upon pro-nuclear interests in Europe. The intensity of the Creys-Malville demonstration in the summer of 1977 showed the extraordinary passion of the European nuclear critics' opposition to the breeder reactor. The implied criticism of breeder reactor development and deployment contained in the new American policy gave enhanced respectability to the anti-breeder, anti-plutonium campaign and, hence, enhanced credibility to nuclear critics.

Less than a year after its formulation, many European government officials saw an important, if unintentional, side-effect to the new American policy:[9] a strengthening of antinuclear forces and, therefore, a new impediment to meeting energy needs with nuclear power.

President Carter's attempts to prevent the proliferation of nuclear weapons seemed to have deepened the stalemate over nuclear power in Western Europe.

It also appeared to have created new problems for the nuclear industry at home.

New Problems for the American Nuclear Industry

Many in the United States shared the European objection to the Carter policy.[10] Ironically, though, a common domestic reaction to it was somewhat at odds with the implications foreseen in Western Europe. To the American nuclear industry, the policy implied an outright concession of decisive technical and commercial advantage to foreign companies and governments. Uranium was a scarce resource and, therefore, the United States could not possibly affect the inevitable introduction of the plutonium economy. Efforts to do so merely meant that the advantages of this important technological innovation would accrue to others, notably to the advanced French and German industries.

Deferring the onset of the plutonium economy will cause several practical problems for the renewed growth of American nuclear generating capacity. The first of the practical difficulties implied by the new American government policy involves spent fuel storage. In the United States spent reactor fuel was piling up in water pools at power plant sites and in similar pools at incomplete reprocessing facilities.[11] Although such storage is safe and inexpensive, it is only temporary.

The Department of Energy has assigned high priority to

a program for solving the spent fuel storage problem. The central premise of this effort is that the problem is not technical in the standard sense. The officials in charge of the government's "radioactive waste isolation" program believe that the technology and know-how is in hand to proceed with a more sophisticated disposal scheme than storing spent fuel assemblies in water pools, provided that some options are retained for response to new data and experience.[12] But they also concede that some important *political* questions remain unanswered. What is *adequate* nuclear waste management and disposal? For how many years *should* container integrity be assured? If leakage does occur, what is an *acceptable* rate? In short: How safe is safe enough?

In principle, a wide variety of standards could be applied to any long-term spent fuel or high level waste repository design to indicate "adequate" safety. Nuclear power plant construction proceeded before the question of "how safe is safe enough?" had been answered to everyone's satisfaction. As we have seen the result was delays, equipment replacement, uncertainty, and cost increases. This experience argues for caution about designing and building a long-term spent fuel or high level waste repository before there exists a broad and explicit consensus on what we want such a facility to accomplish. Unfortunately, however, this whole situation seems to be tailor-made for the extreme antinuclear critics intent on further embarrassing the proponents of nuclear power.

It is difficult for us to avoid the conclusion that the stalemate over nuclear power did indeed deepen during 1977. But even if it did not, it was clear by the end of that year that the increasingly emotional politics of the nuclear safety controversy would have a decisive impact on the role of nuclear fission in most industrialized societies for decades.[13]

CHAPTER 11

The Abuse of a Technology

THE 30-year effort to develop nuclear fission as a cheap and reliable energy resource has been one of the 20th century's most important planned efforts to modify society. The campaign has been controlled by the interaction of four groups: scientists and engineers with an intellectual stake in its success; businessmen with a financial stake in its success; public officials with a political stake in its success; and a heterogeneous array of variously motivated individuals and organizations with a stake in its failure. The outcome has been technological virtuosity combined with commercial and political debacle, resulting in a stalemate. How did it happen? The time has come to draw together the threads we have been following through this dense and difficult thicket.

The Nuclear Scientists and Their Government Patrons

Many of us grew up believing that nuclear fission held revolutionary implications for many human activities. The grounds for this belief were initially articulated at the end of World War II by scientists with a profound intellectual and psychological stake in its ultimate confirmation. It quickly hardened into doctrine that was everywhere accepted as revealed truth. Part of the reason for this acceptance was that no independent social institution was yet in a position to doubt the wisdom of the scientists who simultaneously invented and sanctified the meaning of atomic energy. Also important to American acceptance of this belief were the Joint Committee on Atomic Energy and the Atomic Energy Commission. These new government agencies had an independent organizational stake in sanctifying the prophecies of the nuclear scientists. It was in their political interest to proselytize favorable scientific and engineering judgments. Though radical in detail, the government atomic energy machinery fit the standard American pattern of providing "clientele" interests with public agencies to further their own ends.

What has not been generally appreciated is that the original clients of these two government bodies were the scientists in the national laboratories established as the heart of the wartime atomic bomb development program. Fundamentally, it was the interests of these scientists that the Joint Committee and the Commission protected and promoted. It soon became evident, however, that the national laboratories were unable to proceed with either their civilian or their military tasks without the continuing help of their wartime partners from industry. Development of nuclear technology was an industrial enterprise; it required production as well as research.

A number of American companies, including several who had served as contractors to the army's Manhattan Project, were naturally willing to provide, for modest fees, vital services to the government and the government's scientific clients in the peacetime nuclear development program. It was entirely reasonable for many of these companies to hope that the expertise they gained as contractors to the federal research and development programs could eventually be turned into a profitable hardware business. Their executives could hardly have been able to question the validity of the doctrinal postulates on which the anticipation of such opportunities was based. The executives who undertook to manage the government's nuclear development programs did not do so because of the small fees their companies were paid. They did so, in part, because they believed that the experience would eventually yield important commercial advantages.

The national laboratory scientists and their government patrons were at cross-purposes with industry from the outset. To the laboratory director, nuclear development was preeminently a research problem; to the AEC's business contractors, it was an industrial and marketing problem.

Only in one corner of the AEC's early development program did the interests of the government and its contractors converge: Admiral Rickover's Submarine Propulsion Program. The admiral wanted to produce reliable machinery as quickly as possible. To him, the problem was not to design the most advanced reactor, but to build nuclear-powered boats that worked.[1] His contractors required no daring leap of imagination from a working nuclear-powered submarine to a working nuclear-powered generating station that could be sold at a profit.

As we have seen, the Atomic Energy Commission chose to play a very narrow role in the early stages of light water power plant commercialization. The principal consequence was that when the manufacturers made their historic "turnkey" gamble, the agency had no independent way to judge these developments. At its senior operating levels, the agency was chiefly staffed with contract administrators and engineers whose experience centered on uranium and plutonium production. Nowhere in the AEC was there anyone who could competently assess the commercial situation that was developing. Moreover, the Commission had a strong institutional interest in accepting rather than challenging the manufacturers' economic claims. For more than 10 years the Joint Committee on Atomic Energy had insisted on a more aggressive government reactor development program. Finally, in Oyster Creek and the Great Bandwagon Market, the AEC had dramatic vindication of its bitterly contested power reactor development policy.

In further policy decisions in the aftermath of Oyster Creek, the AEC guaranteed that the government would become even more remote from the world of commercial nuclear power. Just when the light water reactors were first being sold in large numbers to electric utilities, the Commission, which until then had been absorbed by the choice among several advanced reactor concepts, turned its attention to the liquid metal breeder reactor, an entirely new reactor technology with a host of new and different development problems. As a consequence, government resources crucial to light water development were being directed toward problems with little direct relevance to the needs of the light water manufacturers or their new customers. This allocation of administrative, fiscal, and intellectual resources was based on the premise that successful operation of a prototype nuclear plant meant that the whole technology had been mastered.

This belief was not peculiar to Americans. It was shared by government agencies in other countries with similar development responsibilities. In Europe, probably only the relatively limited amount of available money prevented a similar proliferation of pilot projects and various prototype reactors. In most of the countries pursuing nuclear power programs during the early 1960s, it was supposed that the solution to the competitive liabilities of earlier nuclear plants was to develop more efficient and advanced reactors. For 20 years, it was the common belief in the United States and Europe that in order to improve the economic performance of nuclear power plants technicians had essentially to improve the physical performance of the reactors. This conviction grew out of a peculiarly scientific definition of the problem, a definition which allowed a host of unsolved industrial and public-opinion problems to go completely overlooked.

In retrospect, the temptation to concentrate on development of "the next technology" should not be a surprise. After all, that was precisely the job for which the scientists were trained and the national laboratories were founded. But the way "reactor development" was defined and managed in the United States caused a broad chasm to open between the bureaucrats in Washington and the scientists in the laboratories on the one hand and the needs of the companies with a stake in the emerging light water power reactor business on the other. The principal frame of reference for the scientists and administrators was competition for funds to pursue various advanced concepts by building prototypes. The laboratories had their favorite concept, and the object of their game was to persuade the AEC that theirs offered the best chance of economical nuclear power. The AEC was, of course, competent to assess the scientific and engineering propositions in these sales pitches. But it had little competence to make independent economic or business judgments about the commercial prospects, or even the commercial practicality, of the programs which were proposed to it.

The simplistic notion of the relation between improving the physical performance of reactors and improving their commercial status was inadequate in two ways. First, the design of the most advanced concepts often led to difficult engineering problems. Second, although it may have seemed rational to explore alternative design concepts before choosing the "best machine," this strategy completely overlooked the realities of the market for reactors. The potential manufacturers were evidently not interested in waiting for the results of lengthy experimental programs and prototype construction efforts so they could choose the most efficient concept. The public record of their actions suggests that they believed themselves to be under strong pressure to rapidly turn the know-how they had gained as government contractors into a profitable business.

To be fair to the scientific establishment, we should point out that very few atomic scientists apparently believed that the "primitive" light water reactor could ever fulfill the promise of nuclear power.[2] To the scientific community, light water reactors were the accidental by-product of early military experimentation directed at submarine propulsion. It was difficult for most of them to believe that the solutions to the particular problems posed by submarines might also be appropriate for the generation of electricity on land. Moreover they knew that light water reactors were among the least efficient consumers of uranium.

A world-wide nuclear power program seemingly locked into light water systems is a difficult thing to accept for many scientists and engineers who have tried to base commercial nuclear power on rational choice among alternatives—alternatives whose most important apparent advantages were technical. But the real world ignored or overode technical rationality. Military, political, and commercial constraints turned out to have a decisive causal influence on the development of nuclear power.

It is ironic to observe in retrospect that the only technology which was temporarily able to compete with light water was a British design—gas graphite—which was also a by-product of military objectives. The British initially designed and built their gas-graphite plants for plutonium production rather than electricity. In fact, among all of the early attempts to design a reactor to generate electricity, only the Canadian "CANDU" design was motivated from the beginning by nonmilitary objectives. But, the entirely nonmilitary Canadian design was the first to be used surreptitiously—by India—to produce plutonium for a bomb.

In 1965, Milton Shaw accurately predicted the circumstances in which the nuclear power industry would find itself some 10 years later.[3] Unfortunately, his own efforts to rectify the situation he inherited when he became Director of Reactor Development for the AEC simply made matters worse. By redirecting the government program toward development of a breeder reactor, Shaw squandered his own considerable talents and most of the AEC's resources in yet another program with only long-range relevance to the commercial situation. His assignment of a low priority to the Commission's light water safety research program was especially unfortunate.[4]

The Reactor Manufacturers and Their Customers

By the early 1960s, American reactor manufacturers had apparently become persuaded that a commercial market for light water reactors was far more imminent than any informed person would have thought during the 1950s. They also apparently believed that the sooner they were able to take advantage of this potential market, the larger the payoff to them. But, most important, they concluded that by taking a calculated financial risk which was large in absolute dollars but small compared to the expected rewards, they would be able to stimulate this market while simultaneously securing for themselves a guaranteed share of it. Hence the turnkey offers.

Did the manufacturers really believe the early light water reactors would actually produce electricity at costs competitive with coal? We have no evidence that they did not. But we have been able to identify two other beliefs that seemed to have had an important, if not decisive, effect on their actions. First, they believed they could estimate the cost of these plants within a relatively narrow band of uncertainty. When they made the turnkey offers, both General Electric and Westinghouse apparently believed that they were able to estimate the cost of a nuclear plant within 15 percent on the basis of prior experience. Second, they believed they could count on "learning effects" to produce a savings which would inevitably compensate for any losses incurred on the early turnkey offers. Both of these beliefs were serious errors in business judgment.

These errors argue that neither Westinghouse nor General Electric fully comprehended the truly novel nature of the business they were entering. This lack of understanding was probably partly due to everyone's confusion about the relation between successful prototype operation and successful reactor development. Both companies accepted their previous technical success with submarine propulsion reactors and, later, with the first light water reactor prototypes as proof that light water reactor technology was "in hand." Both the manufacturers and their customers overlooked the fact that development of commercially competitive nuclear technology required much more than a safe and reliable power plant. It required the establishment of entirely new and intricately interrelated industrial processes and services. And it required public acceptance.

When General Electric was considering whether to compete with Westinghouse, a member of the General Electric Board of Directors reportedly asked his associates: "Why do we wish to be in the boiler business?" The question is revealing. It was not a profound question. It was the wrong question. A decision to enter the nuclear power business represented far more than a decision by a steam-turbine company to move backward into boiler manufacture. It meant a commitment to develop manufacturing competence in a host of enterprises only slightly related to the company's previous activities. The nuclear power business was only trivially more "natural"—for instance, in the sense that it meant working with large steel structures—for General Electric than it would have been for Du Pont, another wartime contractor of the government.

Business and government both acted as if they forgot or

never clearly understood that, like any high technology product, a power reactor requires the support of many ancillary services and industrial processes. Just as the most advanced television receiver is worthless without a source of electricity, the best nuclear steam-supply system is dependent on supporting services. A crucial mistake of the early American nuclear program was the assumption that private incentives enhanced by limited government financial assistance would make these processes and services automatically available as soon as an effective reactor had been designed. In Europe, only France and, to a lesser extent, Great Britain avoided this same mistake.

The consequences of this blunder are especially evident for the light water fuel cycle. In effect, random attention was given to various segments of this extremely complex and interdependent system, depending only on whether someone happened to perceive, correctly or incorrectly, the possibility of making a profit in developing them. Such patchwork progress led to an extremely uneven development of the components in the system. This failure represents more than a few errors of omission, such as the highly publicized failure to develop a suitable waste-disposal technology. It raises a question about the ability of independent interests pursuing their own goals to design and provide a wide range of services and processes, the performance of each of which is sensitive to the characteristics of each of the others.

The Light Water Innovation Process

We can, therefore, ask sharp questions about the actions of most of the participants in the light water story. The purpose of these questions is not gratuitous, easy, retrospective criticism. Our intent is not to allege mendacity or even foolishness to people who were merely trying to do the best they could with limited means in the service of often admirable, sometimes noble, objectives. Rather, our purpose is to highlight avoidable flaws in the system within which these people operated.

For nearly a quarter of a century the theology of nuclear power—unchallenged and unchallengeable—was accepted by a variety of diverse interests to advance a variety of diverse causes: the protection of American private enterprise against "socialistic" encroachment; the political unification of Western Europe; the economic development of Third World countries; and, of course, as the answer to a greedy OPEC. Rarely did those who seized on nuclear power as a means to their own ends know its actual economic and technical status. Instead, the information available to them was part of a catechism whose basic function was to answer infidels and sustain the faith of the converted. The result, a circular flow of self-congratulatory claims, preserved the discrepancy between promise and performance.

Systematic confusion of expectation with fact, of hope with reality, has been the most characteristic feature of the entire 30-year effort to develop nuclear power.[5] This confusion was unnecessary. It was the unintentional result of consciously designed institutional relations among American and Western European scientists, public administrators, politicians, and business executives.

The identification of promise with performance began in the United States. The economic "analyses" which controlled discussion during the critical early years of light water commercial sales had nothing to do with the detached confrontation of proposition with evidence which we think of as analysis. The public agencies with putative responsibility for facing the facts had neither the means nor the motivation to respond critically to the nuclear industry's propaganda; they could only sanctify it. This they did with notable eagerness.

By the mid-1960s, it was clear that tens of megawatts of light water reactor-generating capacity would be built in the United States in the coming decade. Responsible public officials never seriously questioned whether this was happening in a manner consistent with the protection of other interests. That the pace, scope, and circumstances of light water commercialization were necessary and desirable was always an unquestioned assumption, never a conclusion.

By pursuing institutional interests only distantly related to those of the public at large, the Atomic Energy Commission and the Joint Committee on Atomic Energy became soapboxes for light water reactor promotional literature; indeed, this was their most important role after Oyster Creek. The influence of the Joint Committee on Atomic Energy was especially pernicious, for it, not the President, was the Atomic Energy Commission's effective sovereign. The 1946 atomic energy legislation established a perfectly insulated, self-perpetuating organization with plenary power. In doing so, it virtually ensured many of the events of the ensuing 30 years. The Joint Committee's role in this disaster is a textbook illustration of how to guarantee the triumph of special interest over public interest.[6]

None of these things happened accidentally. The theology of nuclear power and the sanctification of light water technology created an interlocking set of intellectual, political, and commercial interests. Scientists with an intellectual stake in the success of nuclear power, politicians with a political stake, bureaucrats with an organizational stake, and businessmen with a commercial stake reinforced and amplified each other's claims. Much of this misfortune appears to have its roots in the early American mistake of fitting nuclear power development into the client-patron pattern of government. By serving as soapboxes for the economic claims of the reactor manufacturers, the Atomic Energy Commission and the Joint Committee significantly amplified the flow of misinformation and decisively altered the commercial strength of these companies at home and abroad.

In the United States, the AEC's acquiescence in the economic claims of the hardware salesmen was almost certainly as important to the commercial "success" of light water reactors as the various government subsidies to their research efforts. In Europe, the AEC's effect was even more dramatic: it eliminated competitive technology from the field because the European nuclear industry did not have the political power of its American competition. Thus, when falling oil prices during the 1960s began to make the economics of gas-graphite reactors look increasingly dubious, American light water reactor manufacturers could still count on continuing official confirmation of their optimistic expectations. Faced with officially sanctioned prognoses of the economic status of light water reactors, European governments began to have doubts about the wisdom of their own commitment to an apparently inferior technology. Throughout Western Europe, the "turnkey" contracts and the Great Bandwagon Market were accepted as proof of a decisive economic and technical breakthrough. Consistent with popular clichés about American technical superiority, the rush in the United States to light water was naively interpreted in Europe as the signal that electricity producers in the rest of the world would soon do likewise. European utilities and potential reactor manufacturers were eager to be first in line for a share of the benefits and profits of this business. Later, they just as eagerly awaited good news from across the Atlantic to confirm the wisdom and foresight of their actions. There was little incentive to question light water cost experience or performance.

In retrospect many Europeans might be inclined to be bitter, since they have apparently been the victims of the same confusion between expectation and fact that originated on the other side of the Atlantic Ocean. From a European perspective, the worldwide triumph of American nuclear power technology is almost a caricature of American technological imperialism. But these feelings might be tempered by the uncritical manner in which many Western Europeans accepted the American reactors manufacturers' claims, as another proof of their general faith in American technological omnipotence.

CHAPTER 12

The Necessity for Compromise

IT HAS BEEN less than 50 years since the image of a factory with smoke billowing from its stacks was a widely recognized symbol of social progress; today it is a symbol of pollution. The attempt to transform nuclear fission from a scientific discovery into a commercial reality has occurred during the time when the social values shaping such images have been changing most rapidly. At the end of World War II, there was still a consensus in the United States and other industrialized societies about the absolute benefits of scientific and technological progress. This consensus no longer exists. Now different conceptions of what is in society's interest have been articulated and have attracted followings.[1] The effects of these changing values are everywhere apparent.

Suppose someone had certain knowledge that an oil field equivalent to the Alaskan North Slope existed under the Grand Canyon. The ensuing controversy about whether to exploit such a field would dwarf mere technical debate. We would witness—and probably become part of—a war of values centered on disagreement over the relative merits of increased oil supplies and perhaps less costly energy versus the modification of a unique landscape. A similar debate has accompanied construction of the Trans-Alaska oil pipeline.

It has been widely believed that there is a direct relationship between prosperity and energy, with economic growth requiring ever-increasing consumption of energy. But during the 1970s, the apparent implications of a continuously growing demand for energy conflicted with other social goals. Consumption of oil and coal threatened inevitable degradation of the natural environment and hazards to human health and property. For many decades we tacitly accepted these consequences, probably because those most directly affected had little power to do otherwise. Now the partial substitution of uranium for coal and oil seems to offer new kinds of threats to the physical environment and to society, and it heightens political and economic interdependence among countries.

Western society has been shaped by the availability of cheap energy from oil. The benefits of our apparent good fortune were obvious to most persons well before the costs. For example, the undesirable side effects of a transportation system based upon private automobiles did not become easily apparent until the automobile had assumed a central role in our lives. For this reason we now find change extraordinarily difficult and costly. The simple arithmetic of growth makes it so: an economy which grows at an average annual rate of 4 percent will double in size every 18 years. No wonder that a constant growth rate suggests to some accelerating damage to the environment and accelerating depletion of natural resources.

Opposition to nuclear power technology has been only a part of the protest against this general trend, but the circumstances of light water reactor commercialization made this technology an especially vulnerable target of environmental protest. By denying to those outside the "club" the right to question their actions, the proponents of nuclear power in industry and government damaged their own credibility and enhanced the legitimacy and the power of their critics.

Technology Assessment

An obvious lesson of the reactor development and commercialization process in the United States is that those with a stake in the success of a technological innovation must not control the information about it. If they do, it is inevitable that they will distort or obscure the issues of fact and value which the innovation engenders. This principle has to a large degree already been accepted as the foundation for plans for future programs in the United States and Western Europe. Planners now generally accept the need for open, pluralistic, ex ante analysis of the social costs of new technology, even though they have not yet widely implemented their intentions. The fashionable term for this activity is "technology assessment." Like many fundamentally sound ideas, technology assessment is already burdened with an accretion of methodologically and conceptually obscure baggage, much of which is as unnecessary as it is pompous. To really prevent abuses of technical privilege, technology assessment requires little more than the ability to read and count, together with an independence of mind. Basically, the job is to ensure that differing views are articulated and fairly considered before policymakers make any irrevocable decisions.[2]

There has been a considerable effort to institutionalize technology assessment in the American federal government during the past few years. Perhaps the most notable example is Congress's Office of Technology Assessment. This strategically placed organization and similar offices and bureaus in the Executive Branch now give most new claims for research or development funding far more outside review than was usual in the 1960s. As it becomes part of the way Americans deal with innovations, technology assessment promises to remedy the most important flaw brought to light in the reactor commercialization process: the tendency to create a circular flow of mutually reinforcing prophecies insulated from outside criticism during the early stages of development and commercialization.

At the same time, the light water story suggests grounds for caution about the future of technology assessment. One of the most persistent difficulties in evaluating the social acceptability of nuclear power is that its benefits and costs accrue to different groups. At many points in the early commercialization, those who believed themselves in jeopardy from nuclear power were not those most likely to benefit from it. In a decentralized political system, this characteristic increases the practical difficulty of arriving at a society-wide consensus. Furthermore, the lack of a straightforward and generally accepted standard for comparing benefits and costs exacerbates this problem. There is no good way to equate jobs with morbidity or mortality, and many technologies offer economic growth for a measure of sickness and death. Our abiding inability to handle such trade-offs in a convincingly rational manner invites emotional and rhetorical issues into the debate, increasing its difficulty and sometimes its bitterness. Finally, key decisions must inevitably be made with great uncertainty about the relative and absolute magnitude of their costs and benefits. It could very well be that the conventional 15 percent "contingencies" used by many commercial and business interests in developing light water reactors were necessary to avoid immobilization by uncertainties of much greater magnitude. Even if this were not plausible, the drive for technology assessment poses a difficult problem in political accommodation.

The Political Problem

For most of the 20th century, the American democratic process has moved systematically toward greater representation and openness. Many interests which were effectively disenfranchised as few as 20 years ago now have the power to make their grievances heard and their needs felt. Enhanced opportunities for participation in the political process, often through the courts, and greater responsiveness to minority interests are profound developments in American public life. Short of the threat of national catastrophe, it is very unlikely that this trend will reverse in the future. But much of the rhetoric about "reform of the reactor licensing process" assumes just such a reversal. If reform means moving toward more narrow rights of participation in social decision making, it will be much more difficult to accomplish than those who speak of "cleaning up" or "straightening out" the regulatory process believe. What seems to be reform to one group is denial of constitutional rights to another. It is obvious which of these positions is more powerful.

The political challenge may be greater in Europe. It is certainly worth taking full account of the fact that a stalemate over nuclear power has developed in the United States, whose political system has a long history of working out compromises among sharply conflicting objectives. The European political systems, with less inclination and tradition for seeking compromise, are likely to find the nuclear power issue especially difficult and troublesome. This assessment cannot be great comfort to Western political democracies who look to nuclear power as an important part of their solution to their OPEC energy and balance-of-payments problems. The trends in representative democracy on which the American nuclear critics have capitalized appear in general to be common to all of the industrialized West and

Japan. It is reasonable to expect that the political process in all of Western society will become more, not less, open and more, not less representative. Minority interests everywhere will probably enjoy enhanced power—even if it consists only of the negative power of effective veto.

For these reasons, efforts to control technological innovation may very well foreclose some options which seem desirable even to considerable majorities in the community. The Western democracies confront a vexing task in striking a reasonable compromise between closed decision making and a system that requires full accommodation of all minority objections to any positive action.

Some nuclear power critics will not rest until the technology is abandoned. The political democracies' widening commitment to technology assessment may well provide these critics with the means to impose their desires on society. Opening the decision-making process to a variety of outside views is supported, at least superficially, by most people. Whether this means the *de facto* triumph of the nuclear opposition or the final vindication of the nuclear advocates remains to be seen. But, technology assessment may well substitute the tyranny of the extreme antitechnocrat for the absolutism of the technocrat.

The Necessity for Compromise to Resolve the Stalemate

Many of us are disturbed by this reactionary prospect. Nonetheless, it appears to be imminent for nuclear power. The only way to avoid the triumph of the antitechnocrat seems to be further compromise by those who are most persuaded of the need for a significant, long-term energy contribution from atomic fission.

The motivation and the real objectives of many American and European nuclear critics are not completely clear, but a few generalizations seem safe. A difficult thing for Europeans to understand about American nuclear critics is that some of the most prominent and influential are neither political radicals nor even "ecologists," as that label is commonly used in Europe.[3] For them opposition to nuclear power does not go beyond rejection of a specific technology. In the United States, nuclear critics are not always environmentalists, nor are all environmentalists opposed to nuclear power.[4] American opposition to nuclear power crosses traditional ideological and party lines.

To a certain extent the same appears to be true in Europe. But there is an important difference. In Europe, opposition to nuclear power has become a means to achieving more fundamental ends by true radicals—who often label themselves "ecologists"—desiring basic change in the economic and social fabric of their societies.[5] In other words, opposition to nuclear power is one of the most effective rallying points around which European social critics gather.

In this apparent difference between Europe and the United States can be read a possible way to avoid the *de facto* victory of the nuclear critics which the present stalemate threatens. American nuclear proponents may find an acceptable—if not for them ideal—compromise in the separation of the most adamant critics from their less extreme and more pragmatic supporters. By generous interpretation, evolving Carter Administration policy could be construed as an attempt to accomplish this separation. By officially sanctioning the allegations about some of the potential dangers of using plutonium as a reactor fuel, President Carter may have regained the support of those who object to nuclear power on the basis of its implications for nuclear weapons proliferation.

But the American nuclear industry has been unwilling to concede any flaw in its vision. The battle to preserve the original nuclear dream proceeds in the courts and on the bill-boards.

Were the nuclear industry to ask our advice, we would predict that unless the proponents of nuclear power refocus their objectives, they will lose this battle. We know of no evidence that indicates that the nuclear critics are any less successful now than they have been in using the reactor licensing process to impede power plant construction. Indeed, the contrary appears to be true. More and more often, the time-consuming and costly burden of proof has shifted to the electric utility companies to make the case that nuclear power plants are safe, reliable, and economical.

In taking their message to the public at large, the nuclear critics' success has been harder to gauge. Certainly the nuclear industry can justifiably claim victory in the state-wide popular referenda held in 1976. But these undeniable public opinion victories will, in the long run, have less effect than the cumulative, though difficult to measure, impact of the fearsome images which the critics can conjure and exploit.[6]

In brief; the nuclear industry is fighing a battle in an environment congenial to its opposition: continued reliance on adjudicatory proceedings or on public relations campaigns will not produce victory. It seems obvious that the time has come for negotiation and compromise to replace adjudication and advocacy.

A significant area of potential compromise is spent reactor fuel management and radioactive waste isolation. As yet, neither the nuclear industry nor the nuclear critics have staked out full technical positions; at this time they can discuss a number of options without the need to defend prior public commitments. Equally important, few expensive hardware or facility construction commitments have yet been made by either industry or government.

A wide variety of specific technical standards could, in principle, be applied to any nuclear waste management scheme to indicate "adequate" safety. In addition, a large number of technical and economic trade-offs can be made among alternative waste management and isolation schemes. For example, is it more economical to store high level radioactive wastes in water-cooled annular canisters or in air-cooled cylindrical ones?

The waste management problem is urgent enough to hold everyone's attention, but not yet sufficiently acute to require instant action. Moreover, there is an opportunity in solving this problem to attempt what was not done prior to when commitments were made to build the first commercial light water reactors: to develop an explicit consensus among the nuclear critics *as well as* among nuclear advocates on the technical standards which answer the question, "How safe is safe enough?"

We do not suppose that building such a consensus will be easy nor are we confident that it is even possible. Certainly, the nuclear critics have shown little evidence of their willingness to compromise on anything. In order to develop workable standards for radioactive waste management and isolation, they will have to resist the temptation of putting government and the nuclear industry in the impossible position of "demonstrating" something which by definition would take decades or even centuries. Similarly, government and the nuclear industry will have to resist the temptation to rush ahead with a facility construction program in order to claim that "the problem has been solved." Finally, any serious effort to reach consensus on waste management and isolation will require government to fairly represent the opinions of both sides—and to show great skill in mediating between them. This is a lot to ask from all of those involved in the nuclear power controversy.

The economic issue is a possible second area for compromise. Naturally most citizens want the cheapest possible electricity. Government regulators are eager to satisfy this popular demand. Electric utility executives are no less interested in generating power in what seems to them to be the most economical way. It is hardly surprising, therefore, that the economic issue has played a central role in the nuclear controversy. But the information does not exist to resolve the question one way or another and will not exist for several years more. In these circumstances it seems best to proceed on the assumption that all of us interested in cheap electricity can afford, in a narrow microeconomic sense, either nuclear or coal-fired power. Temporarily suspending judgment would remove a contentious issue whose inconclusive nature is merely serving to polarize opinions and exacerbate antagonism.

The same is true of the American nuclear industry's unremitting insistence on the urgency of breeder reactor development and deployment. A third area for compromise is admittedly somewhat one-sided. But, we suspect that the political benefits of letting sleeping breeders lie may be considerable. We have the impression that many persons—particularly many academics and journalists—who might support renewed growth of light water reactor generating capacity balk at the vision of hundreds of breeder reactors and a plutonium-based electricity system.

There may well be technical and economic costs to a nuclear power program which meet the demands of even the less extreme critics. Their support, however, seems to be crucial to *any* renewed growth of nuclear generating capacity in the United States.

A consensus-building effort of this sort may, unfortunately, be less pertinent in Europe because the controversy has a more symbolic nature there. Western European political leaders must deal with a confrontation between different conceptions about the proper structure of modern society and the place of high technology in it. Hence, essentially technical concessions to the critics of nuclear power are less likely to suffice as a solution to the political problem which the critics present to European governments.

The nuclear debate has given ecologists and radical politicians an opportunity to get their message across to more of the European public than they have been able to reach before.[7] Meanwhile, European energy planners have continued to reaffirm their conviction that nuclear power is the only choice to meet Europe's energy needs; a position based on the 30-year-old vision of an inexorable sequence of technical advances beyond light water reactors will guarantee humanity a limitless energy resource.

A different kind of energy policy rhetoric is probably the first step toward political compromise. This means drastic revision of the nuclear dream; technical considerations, regardless of their validity, may have to take second place to acute political needs. Nuclear power may have to be portrayed as only one among many possible long-term solutions to the European energy problem. Government decisions to build light water plants to meet anticipated growth in electricity demand in the next decade or so do not necessarily imply a long-term commitment to the plutonium economy. Such a change in attitude could be an important way for European political leaders to reestablish the credibility which has been damaged during the nuclear controversy. Of course, to be credible words have to be accompanied by actions. The European governments may also have to make concrete concessions on such sensitive issues as the timing of breeder reactor prototype construction, the conditions for fuel reprocessing, and public participation in power plant siting decisions.

In addition, real and expansive research and development programs will have to be established to investigate several currently unrealistic long-term alternatives to the plutonium economy. The money to do this job properly is unfortunately far more limited in countries like Germany, France, or Sweden than it is in the United States. This limitation will probably oblige Europeans to reestablish meaningful cooperative arrangements in an area where—the light water story has highlighted this—they have had little past success.

In both Western Europe and the United States the increasingly acute political problem with respect to nuclear power is to keep some options open. There is a real possibility, particularly in the United States, of losing three decades of technology as persons with very specialized skills are forced to seek employment outside the nuclear power field and as the dim prospects for renewed growth retard the recruitment and training of new scientists, engineers, and technicians. To avoid such a loss those who have worked the hardest to make the nuclear dream a reality will have to concede on some things they deeply believe themselves to be correct about. Otherwise they will see their dream dissolve forever.

NOTES

Introduction

1. Quoted by: Robert Jungk, *Brighter Than a Thousand Suns* (London: Victor Gollancz Ltd., 1958), p. 198.

2. *Proceedings of the National Association of Manufacturers Conference* were reprinted in: U.S. Congress, Joint Committee on Atomic Energy, *Atomic Power and Private Enterprise*, 82nd Congress, 2d Session, 1952. The volume contains a great deal of material on contemporary expectations about the imminent benefits of atomic energy. See also: James R. Newman and Byron S. Miller, *The Control of Atomic Energy: A Study of Its Social, Economic and Political Implications* (New York: McGraw-Hill Book Co., Inc., 1948). Indispensable as reference documents on the history of atomic energy in the United States are: Richard G. Hewlett and Oscar E. Anderson, Jr., *The New World, 1939–1946. A History of the USAEC*, Vol. 1 (University Park, Pa.: Pennsylvania State University Press, 1962); and Richard G. Hewlett and Francis Duncan, *Atomic Shield, 1947–1952. A History of the USAEC*, Vol. 2 (University Park, Pa.: Pennsylvania State University Press, 1969).

3. U.S. Atomic Energy Commission, Division of Industrial Participation, *The Nuclear Industry, 1966*, p. 76. A megawatt (MW) is equivalent to 1,000 kilowatts (kw). A kilowatt is a unit of "power," i.e., energy per unit of time. One kilowatt is equivalent to 1,000 "joules" of energy per second. One "joule" is a tiny amount of energy that would be able to raise the temperature of 0.25 grams of water from 14.5 to 15.5 degrees Celsius.

4. Statistics on world-wide nuclear generating capacity in operation, under construction, or on order can be found in several industry trade journals. Unfortunately, these sources are not always in agreement. An additional source is the *World Energy Supplies Statistical Paper*, Series J, published by the United Nations. The data cited in the text were taken from the mid-February, 1976 issue of *Nuclear News*. According to United Nations' statistics, however, about 78 GW of nuclear capacity had been installed by the end of 1975. A useful overview is available in: *The Future of Nuclear Energy*, A Hudson Europe Special Report prepared by Mary Mauksch (Paris: Hudson Research, Ltd., 1976).

5. *Public Utilities Fortnightly*, *95*, no. 9 (April 24, 1975), p. 19.

6. Alvin M. Weinberg, "The Many Dimensions of Scientific Responsibility," *Science* 177 (July 7, 1972), p. 33.

7. Ralph Nader and John Abbotts, *The Menace of Atomic Energy*, (New York: W. W. Norton & Co., 1977).

Chapter 1

1. *Proceedings of the First United Nations International Conference on the Peaceful Uses of Atomic Energy* (Geneva, August, 1955).

2. Richard G. Hewlett and Francis Duncan, *Atomic Shield 1947–1952. A History of the USAEC*, Vol. 2 (University Park, Pa.: Pennsylvania State University Press, 1969). See also: David E. Lilienthal, *The Journals of David E. Lilienthal: Vol. II, The Atomic Energy Years, 1945–1950* (New York: Harper & Row, 1964); Philip Mullenbach, *Civilian Nuclear Power* (New York: The Twentieth Century Fund, 1963); and John G. Palfrey, "Atomic Energy: A New Experiment in Government-Industry Relations," *Columbia Law Review* (March 1956).

3. The structure of an atom can be compared to the structure of the solar system. At the center, corresponding to the sun, is a nucleus made up of protons and neutrons. These two "subatomic particles" have approximately the same mass, but protons carry an electric charge, while neutrons carry no charge. In orbit around the nucleus are electrons, particles with far less mass than either protons or neutrons, but which also carry an electric charge. The charge of a proton is positive; that of an electron negative and equal to the proton's.

The number of protons in the nucleus of an element is the "atomic number" of that element. Each element has a unique atomic number: for hydrogen—one; for oxygen—eight; for uranium—ninety-two. The atomic *mass* of an element is the sum of its atomic number (the protons) and the number of neutrons in its nucleus. Unlike the one-to-one correspondence between elements and atomic numbers, different atoms of a single element can have different atomic mass—different numbers of neutrons in the nucleus. Atoms of the same element that have different mass are called "isotopes" of that element. There are several different isotopes of the element uranium. Uranium found in the earth's crust is almost entirely (more than 99 percent) composed of atoms with 92 protons and 146 neutrons in the nuclei. The sum of 92 + 146 is 238. Hence, the most prevalent isotope of "natural" uranium is known as "uranium-238," or in standard scientific notation: "^{238}U." The other two naturally occurring isotopes of this element are: ^{234}U and ^{235}U. The fraction of different isotopes present in natural uranium is known quite precisely: ^{234}U 0.006%; ^{235}U 0.714%; ^{238}U 99.28%.

Isotopes of an element behave identically in chemical reactions. But their nuclear characteristics may vary considerably. To understand why, it is necessary to look at the structure of the nucleus a little more closely. Most of us remember from high school that electric particles of the different charges attract one another, but electric particles of the same charge repel each other. How is it, then, that protons, all of which have a positive charge, can stay bound together in the atomic nucleus? Physicists explain this apparent contradiction to the laws that govern electric charges by postulating the existence of a "strong nuclear force," which overcomes the electric force that would normally cause the protons to fly apart. The "strong" force has been likened to a set of powerful springs to which nuclear particles might be attached. To hold the nucleus together these groups have to be compressed. The stability of these springs under compression by the "strong" force varies considerably among certain isotopes. When a neutron which is free of any connection to other subatomic particles collides with the nucleus of an atom of ^{235}U, at first, the free-moving neutron is "captured" by the ^{235}U nucleus, momentarily creating a new uranium isotope: ^{236}U. However, great internal imbalances in this new structure cause the ^{236}U nucleus to split apart—to *fission*—into two lighter fragments. In addition to fission fragments, an average of two to three high-velocity neutrons are also released. If one of these is captured by another ^{235}U nucleus, a second fission occurs and so on until all the ^{235}U in the vicinity of the emitted neutrons is used up. A self-sustaining fission sequence such as this is called a "chain reaction."

When an atom of ^{235}U captures a free-moving neutron and becomes ^{236}U, the nucleus is torn apart with great violence and in the resulting fission fragments and neutrons are propelled outward at enormous speed. About 80 percent of the energy released by a nuclear fission appears as the kinetic energy of the fission fragments and emitted neutrons. To use this energy to generate electricity some practical problems must be overcome. The isotope, ^{235}U, is one of several which is better fissioned by relatively low-velocity neutrons. To sustain a chain reaction fueled by ^{235}U, the high-velocity neutrons produced by the fission of a given atomic nucleus have to be slowed down or "moderated." Elements with low atomic mass numbers usually are good neutron moderators. Ordinary (or "light") water is a convenient substance since it is readily available and can simultaneously be used as a "working fluid" to transfer heat from the fission process to the electrical generating equipment. The direct method is conceptually the most simple. This design calls for the immersion of the fuel-containing structure—"the core"—in water in the bottom of a large steel chamber—"reactor vessel." The hot core is used directly as a high-temperature heat reservoir to vaporize water which rises as steam to the top of the vessel and is piped to a turbine. After it is used to drive the turbine, the steam is condensed back into water and recirculated to the reactor vessel. Such "boiling water reactors" have only one heat circuit. The material in this circuit (water/steam) functions simultaneously as a working fluid to drive the turbines; as a neutron moderator; and as a coolant to prevent the fuel from overheating and melting.

In a second design for light water cooled and moderated systems, "pressurized water reactors," the water in the core is not allowed to boil. Instead it is kept at a pressure of about 2,000 pounds per square inch. (Normal atmospheric pressure at sea level is about 14.5 pounds per square inch.) When water is contained at a high pressure, the temperature at which it boils is significantly higher than the 100°C (212°F) to which we are normally accustomed. A system of pipes called the "primary coolant loop" transports this superhot, pressurized water from the reactor vessel to another piping system called a "heat exchanger." Operating on the same principle as an automobile radiator, the heat exchanger produces steam in a secondary system or "loop" of piping that has no direct contact with the radioactive water in the primary loop.

Although a coolant such as water can be used to carry heat away from the hot fuel in a reactor core, it cannot be used to regulate the number of uranium nuclei which fission per unit of time. To control the rate at which fission takes place, another control mechanism is needed. Certain elements—cadmium, boron, hafnium—readily absorb large quantities of neutrons without becoming unstable. When rods of these materials are inserted into the structure which contains the reactor fuel, they capture neutrons that would ordinarily cause further uranium fissions. The rate of capture can be precisely calculated and controlled. Hence, like the throttle of an automobile, such "control rods" regulate the entire nuclear heat production process.

In all nuclear reactors, ^{239}Pu is produced as a result of a neutron capture of ^{238}U. Because it is the source of a readily fissionable material, the latter is commonly referred to as a "fertile" isotope. ^{239}Pu is a fissile isotope of the element plutonium.

4. In 1977, all of the enriched uranium used to fuel the world's commercial light water reactors was still being produced by "gaseous diffusion" technology. But a major research and development effort in the United States and abroad had brought a newer technology to the verge of commercial success. This was the so-called "gas centrifuge" method. The chief economic attraction of the gas centrifuge method was that it required a tiny fraction of the energy for operation needed by gaseous diffusion facilities.

5. For a highly readable account of the details, see: John McPhee, *The Curve of Binding Energy* (New York: Farrar, Straus and Giroux, 1974).

6. U.S. Congress, *Commercial Nuclear Power in Europe: The Interaction of American Diplomacy with a New Technology.* A committee print prepared for the Subcommittee on National Security Policy and Scientific Developments of the Committee on Foreign Affairs, U.S. House of Representatives; by Warren H. Donnelly, Science Policy Research Division, Congressional Research Service, Library of Congress, U.S. Government Printing Office, December 1972. Although our interpretation of many of the events to be described in the following pages differs in key aspects from that of Donnelly, his report is a thorough and exceptionally well-documented summary of much of the same material. We highly recommend it to anyone desiring further detail and references on these matters. See also: Harold L. Nieburg, *Nuclear Secrecy and Foreign Policy* (Washington, D.C.: Public Affairs Press, 1964); Jaroslave G. Polach, *Euratom: Its Background, Issues and Economic Implications* (Dobbs Ferry, N.Y.: Oceana Publications, Inc., 1964); Arnold Kramish, *The Peaceful Atom in Foreign Policy* (New York: Harper & Row, 1963); Mason Willrich, ed., *Civil Nuclear Power and International Security* (New York: Praeger, 1971).

7. D. Dolfus and J. Rivoire, *A Propos d'Euratom* (Paris: les Productions de Paris, 1959), p. 80.

8. Bertrand Goldschmidt, *Les Rivalités Atomiques* (Paris: Fayard, 1967), p. 211.

9. Ibid., p. 212.

10. Lawrence Scheinman, *Atomic Energy Policy in France Under the Fourth Republic* (Princeton, N.J.: Princeton University Press, 1965), pp. 9–15; see also: J. E. Hodgetts, *Administering the Atom for Peace* (New York: Atherton Press, 1964), pp. 43–50; and "Les Entreprises

Publiques du Secteur de l'Electricité du Gaz et de l'Energie Atomique" (Paris: La Documentation Francaise, 1972).

11. Alain Peyrefitte, *Le Mal Français* (Paris: Plon, 1976), p. 83.

12. Goldschmidt, *Les Rivalités Atomiques*, pp. 190–192.

13. Ibid., p. 202; and Scheinman, *Atomic Energy Policy in France Under the Fourth Republic*, pp. 86–87, 121–128.

14. Bertrand Goldschmidt, *The Atomic Adventure* (Paris: Fayard, 1962), p. 129.

15. Speech given by Michel Debré, May 11, 1956; quoted by Dolfus and Rivoire, *A Propos d'Euratom*, p. 160; see also: André Fontaine, *Le Monde Diplomatique* (March, 1956).

16. Louis Armand letter in *L'Express*, 6 July 1956.

17. L. Armand, F. Etzel, and F. Giordani, *A Target for Euratom*, reprinted in U.S. Congress, Joint Committee on Atomic Energy, *Hearings*, Proposed Euratom Agreements, 85th Congress, 2d Session, 1958, pp. 38–64.

18. Goldschmidt, *Les Rivalités Atomiques*, p. 229.

19. *Proceedings of the 2d United Nations International Conference on the Peaceful Uses of Atomic Energy, Vol. I* (Geneva: September 1958), p. 113.

20. Goldschmidt, *The Atomic Adventure*, p. 111.

21. Louis Armand and Michel Drancourt, *Le Pari Europeén* (Paris: Fayard, 1968).

22. Quoted by Dolfus and Rivoire, *A Propos d'Euratom*, p. 71.

23. Goldschmidt, *Les Rivalités Atomiques*, p. 218.

24. Henry Nau, *National Politics and International Technology: Nuclear Reactor Development in Western Europe* (Baltimore: The Johns Hopkins University Press, 1974), pp. 132–133.

25. Ibid. For a somewhat different account of the origins of Euratom, see: René Foche, *Europe and Technology: A Political View* (Paris: The Atlantic Institute, 1970).

26. Goldschmidt, *Les Rivalités Atomiques*, p. 141.

27. Dolfus and Rivoire, *A Propos d'Euratom*, p. 141.

28. This section is a revised and re-edited version of: Irvin C. Bupp, "Energy Policy Planning in the United States: Ideological BTU's," in *The Energy Syndrome* edited by Leon N. Lindberg (Lexington, Mass.: Lexington Books, D. C. Heath and Company, 1977). Copyright © 1977, D. C. Heath and Company. Reprinted with permission of the publisher.

29. Hewlett and Duncan, *Atomic Shield, 1947–1952*, p. 40.

30. Ibid., p. 417.

31. Ibid., pp. 421–422.

32. Ibid., pp. 514–515.

33. L. R. Hafstad, "Reactors," *Scientific American* 184, no. 4 (April, 1951), p. 43.

34. Harold P. Green and Alan Rosenthal, *Government of the Atom: The Integration of Powers* (New York: Atherton Press, 1963), pp. 253–254.

35. See Mullenbach, *Civilian Nuclear Power*, for a detailed account of these events; also: U. M. Staebler, "Objectives and Summary of USAEC Civilian Power Reactor Programs," in *The Economics of Nuclear Power*, J. Gueran et al., eds. (New York: McGraw-Hill Book Co., Inc., 1957);

and: Herbert S. Marks, "Public Power and Atomic Power Development," *Law and Contemporary Problems* (Durham, N.C.: Duke University School of Law, Winter, 1956), pp. 132–147.

36. U.S. Congress, Joint Committee on Atomic Energy, *Review of the International Atomic Policies and Programs of the United States*, 6 vols., 86th Congress, 2d Session, 1960.

37. Strauss' own account of these events can be found in his autobiography: *Men and Decisions* (New York: Doubleday & Co., Inc., 1962).

38. Nau, *National Politics and International Technology*, p. 138.

39. Ibid.

40. Ibid. Our own interviews with officials of KRB in April, 1975 substantially confirmed Nau's assessment.

41. Wolfgang D. Muller, "Broad Development in Germany," *Nuclear Industry* 11, no. 8 (August, 1964), p. 26.

42. Goldschmidt, *The Atomic Adventure*, p. 143.

43. Scheinman, *Atomic Policy in France Under the Fourth Republic*, pp. 149–150.

44. Interview with Pierre Chatenet and several other French atomic scientists and government officials who were involved in these matters, April, 1975.

45. Interview with Pierre Chatenet, April, 1975.

Chapter 2

1. "The Jersey Central Report," *Atomic Industrial Forum Memo*, Vol. 11, no. 3 (March, 1964), p. 3.

2. The expression "20¢/mbtu" (twenty cents per million British thermal units) is one of the many ways in which the price of energy can be stated. One btu is equivalent to the quantity of thermal energy (heat) needed to raise the temperature of a standard pound of water from 63 to 64°F. One btu is equal to 1,060 joules. Since coal is employed by the electric utility industry to boil water, its cost has traditionally been expressed in ¢/mbtu, as well as in dollars per ton.

3. "The Jersey Central Report," p. 4.

4. Ibid., p. 7.

5. The expression "mills/kwh" (mills per kilowatt-hour) is the way in which the electric utility industry states the value of a standard unit of electric energy. One mill is one-tenth of one cent ($0.001). One kilowatt-hour is the amount of energy that a one-kilowatt (1-kw) generator would produce if run nonstop for one hour.

Charges on a monthly electric bill include two major costs—generating costs and distribution costs. The former are sometimes referred to as "Busbar costs." The "Busbar" is the location in an electric power system which separates the generating phase from the distribution phase. Throughout this book, the cost of electric energy, usually expressed in mills/kwh, always refers to the "Busbar cost." Distribution costs are those which accrue to operating the network of power lines and transformers which bring electric energy from the Busbar to ultimate users. From the standpoint of comparative economics, the choice of either nuclear generation or fossil generation has no effect on distribution costs. Thus, for the purpose of comparing nuclear power plants with fossil-fueled power plants, distribution costs can be disregarded.

6. U.S. Atomic Energy Commission, *Civilian Nuclear Power—A Report to the President—1962*. Reprinted in U.S. Congress, Joint Committee on Atomic Energy, *Nuclear Power Economics—1962 through 1967*, 90th Congress, 2d Session (February, 1968), pp. 92–253. The claim about the cost reductions since 1958 can be found on p. 130. The "threshold" statement appears on p. 134.

7. U.S. Congress, Joint Committee on Atomic Energy, *Nuclear Power Economics*—1962 through 1967, 90th Congress, 2nd Session, (February, 1968), p. 82. This highly interesting collection documents some five years of exchanges among Philip Sporn, the staff of the Joint Committee on Atomic Energy, the staff and members of the Atomic Energy Commission, the principal reactor manufacturers, and others in the "nuclear community."

8. Ibid.

9. Ibid., p. 83. See also Sporn's 1964 statement: "A Post Oyster Creek Evaluation of Nuclear Electric Generation," pp. 41–56.

10. Ibid., p. 90; letter from George White, General Manager, General Electric Co., to Congressman Melvin Price.

11. Ibid., p. 86.

12. The phrase is Philip Sporn's. It first appears in his letter to the Joint Committee on Atomic Energy, dated December 28, 1967; "Nuclear Power Economics—Analysis and Comments—1967," reprinted, ibid., pp. 2–22.

13. *Nuclear News* 11, no. 1 (January, 1968), p. 38.

14. Ibid.

15. Weinberg's statement is cited by Sporn, *Nuclear Power Economics —1962 through 1967*, p. 5. Sporn's comment can be found on p. 7.

16. The following account draws heavily upon: Irvin C. Bupp, "Priorities in Nuclear Technology: Program Prosperity and Decay in the United States Atomic Energy Commission, 1956—1971," (Ph.D. thesis, Harvard University, 1971), pp. 136–153.

A more complete set of references to the United States Atomic Energy Commission documents and interviews with government and business officials upon which the account is based can be found there. However, the agreement between the author and the office of the Secretary, USAEC, which formed the basis for the original research stipulated that citation of government documents should not be taken to imply that all of them are unclassified.

17. Milton Shaw made this particular argument at a meeting with the Chairman and members of the Atomic Energy Commission, held on April 6, 1965. This was one of several lengthy discussions of the matter by the participants in the period March-April 1965. Complete records of these meetings were maintained by the Office of the Secretary, USAEC. Specific references to times and dates can be found in Bupp, "Priorities in Nuclear Technology." Much of the official record of these events was originally written by one of the authors (Bupp) as a member

of the Secretariat of the AEC from 1963 to 1966. The central argument of this section—that the AEC's reactor development priorities during the crucial 1964–1965 period had little to do with light water reactors—is largely based on that experience.

18. Seaborg's concern as well as the technical and economic rationale for it are reflected in "Summary of Geneva Conference," *Atom*, no. 97 (November, 1964), pp. 218–223. See Bupp, "Priorities in Nuclear Technology," for specific citations to the meetings of the Atomic Energy Commission at which these discussions took place.

19. This exchange and the subsequent AEC decision occurred on April 15, 1965. A public record of the policy debate described in the foregoing pages can be found in the transcripts and supplementary documents of the annual Atomic Energy Commission's "Authorization Hearings" before the Joint Commission on Atomic Energy. See especially: U.S. Congress, *AEC Authorizing Legislation Fiscal Year 1968*, Hearings before the Joint Committee on Atomic Energy, 90th Congress, 1st Session (March 14–15, 1967); Part 2, pp. 660–662, 667–900, and passim.

Much of the High Temperature Gas Reactor story is recounted in: U.S. Congress, *AEC Authorizing Legislation Fiscal Year 1966*, Hearings before the Joint Committee on Atomic Energy, 89th Congress, 1st Session (March 11, 18, 19, 24, and April 13, 1965); Part 3, passim.

The most striking impression which the hundreds of pages of transcript and documents from these legislative hearings conveys is one of self-congratulatory confidence that the light water job had been done. From early 1964 (authorization hearings for Fiscal Year 1965) onwards, the voluminous public record of JCAE hearings reflects a government research and development policy absorbed with the problems of the *next* reactor technology.

Chapter 3

1. The definitive account of the international petroleum market in the 1950's and 1960's is: Morris Adelman, *The World Petroleum Market* (Baltimore: The Johns Hopkins University Press, 1972); see also: Jean-Marie Chevalier, *The New Oil Stakes* (Paris: Calmann-Lévy, 1975, translated by Penguin Books Ltd., 1975).

2. Michel Grenon, *Ce Monde Affamé d'Energie* (Paris: R. Laffond, 1973).

3. For details on oil policy in Europe see Michel Grenon, *Ce Monde Affamé d'Energie*. See also: Chevalier, *The New Oil Stakes*; and Gerard Pile and Alain Cubertafond, *Pétrole, le Vrai Dossier* (Paris: Presses de la Cité, 1975).

4. Michel Grenon, *Pour une Politique de l'Energie* (Paris: Marabout Université, 1972), chapter 2.

5. Goldschmidt, *L'Aventure Atomique*, p. 114. An extremely vivid account of the windscale accident can be found in John G. Fuller, *We Almost Lost Detroit* (New York: Reader's Digest Press, 1975).

6. *The Nuclear Power Programme* (London: Her Majesty's Stationery

Office, June, 1960 (Cmd. 1083)); see also, "The Nuclear Power Programme," *Nuclear Engineering* (July, 1960), p. 316.

7. For a concise survey of the history of the British nuclear power program see: "The Problems and Prospects of Nuclear Power," speech by Sir Christopher Hinton, Chairman of the Central Electricity Generating Board, reprinted in *Atom*, no. 77 (March, 1963), pp. 62–71.

8. Ibid., p. 65.

9. Ibid., p. 67.

10. Ibid., p. 63. See also: British Nuclear Forum, papers for the 5th Foratom Congress, Florence, 1973; and: Foratom, *The Nuclear Power Industry in Europe*, 2nd ed. (Deutsches Atomforum, 1974) and: Mary Goldring, "Britain's New Program," *Nuclear Industry* 11, no. 8 (August, 1964), p. 13.

11. Nicholas Vichney, "The French Program," *Nuclear Industry* 11, no. 8 (August, 1964), p. 23; see also: *Nuclear Engineering* (May, 1964), pp. 158–159.

12. Rapport de la Commission Consultative pour la Production d'Electricité d'Origine Nucléaire ("Commission PEON"), in *Les Dossiers de l'Energie*, Vol. 1 (Paris: Ministère de l'Industrie et de la Recherche, 1963), p. 110.

13. Ibid., p. 64.

14. Interviews with EDF executives, April, 1975 and July, 1976.

15. Jean-Marie Martin, "Planification et Société, Actes du Colloque d'Uriage," in *Etat et Entreprises Energétiques* (Grenoble: Presses Universitaires de Grenoble, 1974), pp. 121–140.

16. Speech given by M. Boiteux at a meeting of the Association des Cadres Dirigeants de l'Industrie, Paris, March 16, 1970.

17. *Nuclear Industry* 11, no. 8 (August, 1964), p. 22.

18. Our interviews with EDF and CEA officials confirmed the existence of a formal but confidential "peace treaty" that was concluded between the EDF president, Gaspard, and the administrator-general of the CEA, Guillaumat. This agreement improved working relations between the two agencies, but it failed to resolve the deep cleavage between them on basic nuclear policy issues.

Throughout this chapter, our account of the relationship betwen EDF and CEA is based upon the personal experiences of one of the authors (Derian) as well as the interviews conducted during our joint research. See also: Dominique Saumon and Louis Puiseux, "Actors and Decisions in French Energy Policy," in *The Energy Syndrome*, Leon N. Lindberg, ed., (Lexington, Mass.: Lexington Books, D. C. Heath and Company, 1977).

19. Interview with J. Horowitz, former chief of the Division of Reactor Development of the French Atomic Energy Commission, April, 1975.

20. Grenon, *Pour une Politique de l'énergie*, p. 332.

21. Interviews with EDF and CEA officials, April, 1975 and June, 1976.

22. *Les Dossiers de l'Energie*, Vol. I, p. 57.

23. Boiteux, speech.

24. *Les Dossiers de l'Energie*, Vol. I, p. 137.

25. Grenon, *Pour une Politique de l'Energie*, p. 335.

26. *Les Dossiers de l'Energie*, Vol. II, p. 10.

Chapter 4

1. U.S. Atomic Energy Commission, Office of Planning and Analysis, *Nuclear Power Growth, 1974–2000*, WASH-1139(74), (Washington, D.C.: U.S. Government Printing Office, February, 1974).

2. The authors' own attempt in 1974 to analyze light water reactor costs using statistical techniques applied to a constant dollar base was the first such systematic analysis of which we were then or are now aware. See I. C. Bupp et al. "Trends in Light Water Capital Costs in the U.S.: Causes and Consequences," Massachusetts Institute of Technology, Center for Policy Alternatives, (MIT: CPA 74–8), December, 1974.

3. The identification of economic promise with fact is the central theme of the Atomic Energy Commission's testimony to the Joint Committee on Atomic Energy in the spring of 1965. See: *AEC Authorizing Legislation Fiscal Year 1966*, passim.

4. Interviews with executives of Westinghouse and Westinghouse licensees in France and Germany, April, 1975 and July, 1976.

5. David L. Bodde, "Regulation and Technical Evolution: A Study of the Nuclear Steam Supply System and the Commercial Jet Engine," (Doctoral Thesis, Graduate School of Business Administration, Harvard University, 1975). See also: Bruce T. Allen and Arie Melnik, "Economics of the Power Reactor Industry," *Quarterly Review of Economics and Business* 3 (1970), pp. 69–79; see also: S. B. Palmeter, "Comparison of USA Reactor Construction Costs—Why so Different?" (Paper presented before the American Nuclear Society 1974 Winter Meeting Special Session "Reactor Construction Management," Washington, D.C., October, 1974); and, John D. Peters, "Nucler Power Plant Construction" (Paper presented at the same conference); and, F. C. Olds, "Power Plant Capital Costs Going Out of Sight," *Power Engineering*, August, 1974, pp. 36–43.

6. For a thoughtful discussion of the public vs. private power issue see: Philip Sporn, *The Social Organization of Electric Power Supply in Modern Societies*, (Cambridge, Mass.: MIT Press, 1971).

7. Harvard Business School, "The Electric Utility Industry: Reference Note" (9–675–170); Distributed by the Intercollegiate Case Clearing House, Soldiers Field, Boston. See also: Michael Tennican et al., *Steam Electric Power Plants: Economic Analysis of Effluent Guidelines* (Washington, D.C.: U.S. Environmental Protection Agency, December, 1974 [EPA–230/2–74–006]).

8. U.S. Congress, Senate, Committee on Interior and Insular Affairs, *Electric Utility Policy Issues*, 93rd Congress, 2nd Session, 1974.

9. The Atomic Energy Commission's report: *Civilian Nuclear Power—The 1967 Supplement to the 1962 Report to the President* (reprinted in *Nuclear Power Economics—1962 through 1967*), is an excellent example. The AEC's discussion of "nuclear economics" ends with an array of misleading bromides: "Capital costs per kilowatt . . . have recently remained fairly constant . . . or have increased slightly. Design innovations of a product improvement nature continue to be incorporated in succeeding plants, and it probably will be a few years before plant pricing stabilizes, though there has already been a move to standard sizes. Currently, there is a substantial back-log of plants on

order. This is stretching out delivery times or presumably has an effect on plant pricing. However, warranted fuel cycle costs have decreased significantly, leading to an overall reduction in the anticipated cost of nuclear energy." p. 278. The final sentence manages to confuse costs with prices while at the same time obscuring the crucial distinction between facts and expectations.

10. The cost data cited in this section were collected in the summer of 1974 for the study by Bupp et al., "Trends in Light Water Reactor Capital Costs in the U.S." They were provided to our research team by the electric utilities, the Federal Power Commission, and the Atomic Energy Commission. One of the early revelations of this research was that these sources often differed among themselves on the cost of a given nuclear power plant.

In 1976 a team of researchers at RAND, headed by William E. Mooz, undertook a replication of our work. Their preliminary results confirm and expand upon the themes discussed here. For additional data on nuclear power plant costs see: William E. Mooz, "Cost Estimating Relationships for Light Water Reactors," A Working Note prepared for the Energy Research and Development Administration, RAND Corp., WN–9926–ERDA, August, 1977; and: United States Atomic Energy Commission, Division of Reactor Development and Technology, *Power Plant Capital Costs Current Trends and Sensitivity to Economic Parameters* (October, 1974); and United States Energy Research and Development Administration, Division of Nuclear Research and Applications, *Nuclear Power Program, Information and Data* (August, 1976).

11. Bupp et al., "Trends in Light Water Reactor Capital Costs in the U.S."

12. Ibid. The point is confirmed by Mooz, "Cost Estimating Relationships for Light Water Reactors." This very careful and thorough research has also uncovered the first convincing evidence of learning effects in the nuclear power plant construction business. "Both [increases in capital costs and construction time] are being reduced . . . as an architect-engineer gains experience in the design and construction of these plants. The experience is particular to each architect-engineer and is not transferred between them. . . . a doubling of the number of reactors [built by an architect-engineer] results in a 5 percent reduction in both construction time and capital cost." p. v.

13. U.S. Congress, *Nuclear Power and Related Energy Problems—1968 through 1970*, Report of the Joint Committee on Atomic Energy, 92d Congress, 1st Session, December, 1971, p. 1.

14. Ibid., p. 30; see also pp. 884–886.

15. Ibid., p. 2. This particular company's problems in producing its first pressure vessels for light water nuclear steam supply systems, as recounted by Harold B. Meyers, "The Great Nuclear Fizzle at Old B & W," *Fortune* (November, 1969), has been used for several years in the MBA program at the Harvard Graduate School of Business Administration as a classic instance of a manufacturing disaster. The 1,000-odd pages of testimony and supporting material compiled in 1971 by the Joint Committee on Atomic Energy about "Nuclear Power and Related Problems" are notably silent on the subject.

16. In our interviews with electric utility company and reactor manufacturer executives, a constant refrain in response to questions about

the apparently disappointing capital cost story of nuclear power was the citation of what they believed to be similar trends for coal-fired plants. An example can be found in a speech given by Gordon R. Corey, Vice Chairman of Commonwealth Edison Co., at the Annual Conference of the Financial Analysts Federation, Washington, D.C., May, 1973, titled: "Nuclear Power; A Successful Technology."

The lack of reliable data together with the wide range of variation in coal-fired plant costs made it difficult to challenge these beliefs until our 1974 study documented that nuclear plant costs had increased more rapidly in the past five years than had those of coal-fired plants.

Unfortunately, almost four years after that study was completed, it is impossible to be more precise about the recent evolution of *relative* capital cost trends because of a persistent lack of information on the costs of building and operating pollution control equipment for coal-fired plants.

We return to this subject in chapter 9.

17. JCAE, *Nuclear Power Economics—1962 through 1967*, p. 9. With considerably understated irony, Sporn also said that "The Browns Ferry 1 and 2 [economic] performance is in a class by itself not explainable at all by conventional economics."

18. U.S. Congress, *Nuclear Power and Related Energy Problems*, esp. pp. 87–154.

19. H. E. Van, M. J. Whitman, and H. I. Bowers, "Factors Affecting the Historical and Projected Capital Costs of Nuclear Plants in the USA," *Proceedings of the Fourth International Conference on the Peaceful Applications of Atomic Energy*, (Geneva, September, 1971), Vol. 2, pp. 427–442.

20. W. K. Davis et al., "U.S. Light Water Reactors: Present Status and Future Prospects," *Proceedings of the Fourth International Conference on the Peaceful Applications of Atomic Energy*, (Geneva, September, 1971), Vol. 2, pp. 21–43.

21. Ibid.

22. *Nuclear Engineering International* (April, 1976 supplement), pp. 14–28.

23. Ibid.

24. Creusot-Loire was created in a major reorganization of the Schneider Group in 1970. Framatome, a subsidiary of the Schneider group, controlled 51 percent of the new company's shares and Westinghouse, 49 percent. In 1976, Westinghouse sold 30 percent of Creusot-Loire's shares to the French Atomic Energy Commission.

"When Framatome received orders for the Fessenheim and Bugey plants, the company employed only 200 persons. It had grown to 887 by 1973 . . . [and] 2,840 by the end of 1976. . . . EDF's Equipment Division had placed its nuclear plant orders *with an industry that did not exist.*" (emphasis added) Michel Herblay, *Les Hommes du Fleuve et de l'Atome*, (Paris: La Pensée Universelle, 1977), p. 225.

25. *Nuclear Engineering International* (September, 1972), p. 705.

26. According to the Commission of the European Community: "With one exception, all community suppliers of light water reactors have to rely to a greater or lesser extent on the technology of U.S. companies. . . ."; see: *Second Illustrative Program for the Community* (Brussels: July 1972, XVII/341/2/71-E), p. 47.

27. Interviews with officials of EDF and CEA, April, 1975 and June, 1976.
28. *Les Dossiers de l'Energie*, Vol. I, p. 110–13.
29. Commission of the European Community, *Second Illustrative Program for the Community*, p. 28.
30. Ibid., Appendix III, "Light Water Power Plants," p. 52.

Chapter 5

1. "Sporn Sees Little Progress in Nuclear Power Economics," *Nuclear Industry* 15, no. 3 (March, 1968), p. 14.
2. Hearings before the Public Utilities Control Authority, State of Connecticut, Hartford, Conn. (February, 1976).
3. For a provocative collection of essays and data on the consequences of the 1973 OPEC actions see: Raymond Vernon, ed., "The Oil Crisis: In Perspective," *Daedalus* (Fall, 1975); see also: John M. Blair, *The Control of Oil* (New York: Pantheon Books, 1976).
4. Executive Office of the President, Council on Wage and Price Stability, *A Study of Coal Prices*, Staff Report, March, 1976: see also: Foster Associates, Inc., *Energy Prices 1960–1973*, A Report to the Energy Policy Project of the Ford Foundation (Cambridge, Mass.: Ballinger, 1974).
5. U.S. Congress, Committee on Public Works, United States Senate, *Air Quality and Stationary Source Emission Controls*, A Report by the Committee on Natural Resources of the National Academy of Sciences, the National Academy of Engineering, and the National Research Council, 94th Congress, 1st Session (Washington, D.C.: U.S. Government Printing Office, March, 1975).
6. Executive Office of the President, Council on Wage and Price Stability, *A Study of Coal Prices*.
7. Arthur D. Little, Inc., *A Study of Base Load Alternatives for the Northeast Utilities System: 1981–1984* (1973). See also: Irvin C. Bupp, "Rationale for Nuclear Power: The ADL Study," (Harvard Business School Case, No. 9–675–106, Cambridge, Mass.), p. 31.
8. As quoted in "Another Landmark Study of Alternatives Favors Nuclear," *Nuclear Industry* 16, no. 9 (September, 1973), p. 28.
9. Ibid.
10. Arthur D. Little, Inc., *A Study of Base Load Alternatives for the Northeast Utilities System: 1981–84*, p. 59.
11. Arthur D. Little, Inc., *A Report to the Long Island Lighting Company on Baseload Alternatives for the 1981–1984 Period* (1975), pp. 1–5.
12. *Proceedings of the First Conference of the European Nuclear Society*, Paris, April, 1975.
13. United States Federal Energy Administration, Project Independence Report (Washington, D.C.: U.S. Government Printing Office, 1974). See also: The Energy Policy Project of the Ford Foundation, *A Time to Choose: America's Energy Future*, Final Report (Cambridge, Mass.: Ballinger, 1974); and U.S. Congress, Joint Committee on Atomic Energy, *Certain Background Information for Consideration When Evalu-*

ating the National Energy Dilemma, 93rd Congress, 1st Session, 1973.

14. See, for example, the interview of Marcel Boiteux, the general manager of EDF, in *La Vie Française* (December, 1973), pp. 170–171.

15. H. Erdemie and B. Réal, *Internationalization dans la Branche "Construction de Centrales Electriques Notamment Nucléaires,"* IREP. Université des Sciences Sociales, Grenoble, October, 1974.

16. Interviews with executives of Creusot-Loire and KWU, April, 1975.

Chapter 6

1. This argument has been widely developed in the French press. See for example, *La Vie Française* (December, 1973) and *La Revue Française de l'Energie* (April, 1974). For an insightful analysis of "pre-OPEC" French energy policy, see: Michel Grenon, *Pour Une Politique de l'Energie* (Paris: Marabout Université, 1972).

2. Stanley Hoffman et al., *In Search of France* (Cambridge: Harvard University Press, 1963); see also: John H. McArthur and Bruce R. Scott, *Industrial Planning in France* (Cambridge: Division of Research Graduate School of Business Administration, Harvard University, 1969).

3. A. Peyrefitte, *Le Mal Français* (Paris: Plon, 1976), passim.

4. Electricité De France, *Resultats Techniques de l'Exploitation* (Paris: Centre De Documentation, Unité Documentaire Murat-Messine, 2 Rue Louis Murat, 75008 Paris, 1977).

5. In April 1975 the authors discussed these matters with several officials of EDF. They were uniformly confident that EDF engineers and project managers would be able to avoid the problems which had caused unanticipated cost increases in other nuclear power plant construction programs.

6. Syndicat CFDT de l'Energie Atomique–*L'électronucléaire en France* (Paris: Editions du Seuil, 1975).

7. See, for example, *Sud Ouest* (11 and 12 November 1974).

8. "Rapport sur la situation de l'énergie en France par Claude Coulais, député," Annexe au procès verbal de la Séance du 5 November 1974 de l'Assemblée Nationale.

9. *Le Point, no. 110* (28 October 1974).

10. For further details on the intragovernmental debate, see: André Oudiz, *Choix des Sites Nucléaires*, rapport ADISH-IREP (May, 1977).

11. *Le Monde*, 4 December 1974.

12. *Le Monde*, 11 February 1975.

13. *Alternative au Nucléaire* (Grenoble: Presses Universitaires de Grenoble, February, 1975).

14. *Le Monde*, 8 April 1975.

15. *Aspects Administratifs de la Régionalisation*, Institut Français des Sciences Administratives, no. 10, CUJAS (1974).

16. Oudiz, *Choix des Sites Nucléaires*, pp. 87–97.

17. *Le Monde*, 5 April 1975.

18. *Le Monde*, 16 May 1975.

19. *Le Monde*, 13 May 1975.

20. The most dramatic events occurred in June 1976 when local protesters destroyed EDF equipment at Plogoff and, for a few days, transformed the proposed plant site into a defensive camp, preventing access by EDF technicians. See: *Le Télégramme de Brest*, 11 June 1976 on the initiative of scientists from the Orsay University and the Collége de France.

21. The "Groupement des Scientifiques pour l'Information sur l'Energie Nucléaire" was formally established in November 1976.

22. *Le Progres*, 9 July 1976.

23. *L'Alsace*, 17 October 1976.

24. *L'Alsace*, 11 February 1977.

25. *Ouest France*, 10 October 1976; 12 April 1977; *Le Républicain Lorrain*, 12 June 1977.

26. *Le Monde*, 28 July 1977.

27. *The International Herald Tribune*, 1 August 1977.

28. John Walsh, "Nuclear Power: France Forges Ahead on Ambitious Plan Despite Critics," *Science* (July 23, 1976), p. 305.

Chapter 7

1. A. R. Tamplin and T. B. Cochran, "Radiation Standards for Hot Particles: A Report on the Inadequacy of Existing Radiation Protection Standards Related to Internal Exposure of Man to Insoluble Particles of Plutonium and Other Alpha-emitting Hot Particles," Petition to the AEC and EPA, Natural Resources Defense Council, Inc., Washington, D.C., 1974. For an especially lucid discussion of all of the alleged hazards of nuclear power see: Royal Commission on Environmental Pollution, Sixth Report, *Nuclear Power and the Environment* (London: Her Majesty's Stationery Office, September, 1976).

2. U.S. Atomic Energy Commission, "Reactor Safety Study—An Assessment of Accident Risks in U.S. Commercial Power Plants" (WASH–1400/NUREG), October 1, 1975. See also: H. W. Lewis et al., "Report to the American Physical Society by the Study Group on Light Water Reactor Safety," April, 1975; and Henry W. Kendall, *Nuclear Power Risks: A Review of the Report of the American Physical Society's Study Group on Light Water Reactor Safety* (Cambridge, Mass.: Union of Concerned Scientists, August, 1975). Finally, see: Joel Yellin, "The NRC's Reactor Safety Study," *Bell Journal of Economics*, Vol. 7, no. 1 (Spring, 1976), pp. 317–339.

3. *The Risks of Nuclear Power Reactors: A Review of the NCR Reactor Safety Study, WASH–1400* (Cambridge, Mass.: The Union of Concerned Scientists, August, 1977).

4. For a summary of the relative accident potentialities estimated by the Rasmussen team, see: "Final Rasmussen Report," *Nuclear Industry* 18, no. 11 (November, 1975), pp. 3–5. See also: Frank von Hippel, "A Perspective on the Debate," *The Bulletin of Atomic Scientists* 31, no. 7 (September, 1975), pp. 37–40.

5. See: 10 Code of Federal Regulations 50.

6. For a thorough technical discussion, see: L. Charles Hebel et al.,

Report to the American Physical Society by the Study Group on Nuclear Fuel Cycles and Waste Management (1977).

7. Arthur S. Kubo and David J. Rose, "Disposal of Nuclear Wastes," *Science 182* (21 December 1973), pp. 1205–1211.

8. W. J. Blair and R. C. Thompson, "Plutonium: Biomedical Research," *Science* 183 (February 22, 1974), pp. 715–721; see also: Donald P. Geesaman, "Plutonium and the Energy Decision," in *The Energy Crisis*, R. S. Lewis and B. I. Stinrod, eds., Bulletin of the Atomic Scientists Publication, 1972.

9. Ibid.

10. For example, Theodore B. Taylor and Mason Willrich, *Nuclear Thefts: Risks and Safeguards* (Cambridge, Mass.: Ballinger Press, 1975); John McPhee, *The Curve of Binding Energy* (New York: Farrar, Straus, and Giroux, 1974). This book originally appeared in serialized form in the *New Yorker*.

11. McPhee, p. 194.

12. T. Greenwood, H. Feireson, and T. Taylor, *Nuclear Proliferation* (New York: The Council on Foreign Relations, 1977).

13. Ralph Nader and John Abbotts, *The Menace of Atomic Energy* (New York: Norton and Co., 1977); McKinley C. Olsen, *Unacceptable Risk* (New York: Bantam Books, 1976).

Chapter 8

1. For an account of one such controversy see: Dorothy Nelkin, *Nuclear Power and Its Critics: The Cayuga Lake Controversy* (Ithaca, N.Y.: Cornell University Press, 1971). For an extensive bibliography on the power plant siting controversy in the late 1960s and early 1970s, see: Dennis W. Ducsik, "Reaching Power Plant Siting Decisions with Environmental and Social Consequences: A New Approach" (M.S. thesis, Massachusetts Institute of Technology, January, 1975); see also: Harold P. Green, "Public Participation in Nuclear Plant Licensing: The Great Delusion," *William and Mary Law Review* 15, no. 3 (Spring, 1974), pp. 503–525.

2. Calvert Cliffs' Coordinating Committee v. USAEC, 449 F.2d 1109 (1971). For an especially thorough analysis, see: Arthur W. Murphy, "The National Environmental Policy Act and the Licensing Process: Environmentalist Magna Carta or Agency Coup de Grace?", *Columbia Law Review* 72, no. 6 (October, 1972), pp. 963–1007; see also Frederick R. Anderson, *NEPA in the Courts: A Legal Analysis of the National Environmental Policy Act*, (Baltimore: Johns Hopkins University Press, 1973), esp. Chapter VII; see also: Izaak Walton League of America v. Schlesinger, 337 Fed. Supp. 287 (1971); and: U.S. Congress, Committee on Interior and Insular Affairs, U.S. Senate, *Effect of Calvert Cliffs and Other Court Decisions Upon Nuclear Power in the United States*, 92d. Congress, 2d Session, 1972.

3. Fred C. Finlayson, "Nuclear Reactor Safety: A View from the

Outside," *The Bulletin of the Atomic Scientists* 31, no. 7 (September, 1975).

4. Ibid.

5. "Criteria for ECCS for Light Water Reactors," *Federal Register* (29 June 1971). For the UCS challenge see: Ian A. Forbes et al., *Nuclear Reactor Safety: An Evaluation of New Evidence* (Cambridge, Mass.: The Union of Concerned Scientists, July, 1971); see also: Daniel F. Ford et al., *A Critique of the New AEC Design Criteria for Reactor Safety Systems* (Cambridge, Mass.: The Union of Concerned Scientists, October, 1971).

6. For a lively, but highly sympathetic, account of the UCS battle with the AEC over "emergency core cooling," see: Joel Primack and Frank von Hipple, "Challenging the Atomic Energy Commission on Reactor Safety," in *Advice and Dissent: Scientists in the Public Arena* (New York: New American Library, 1974), pp. 208–235.

7. A. Wohlstetter et al., *Moving Toward Life in a Nuclear Armed Crowd?* (Los Angeles: Pan Heuristics, April, 1976); Congress of the United States, Office of Technology Assessment, *Nuclear Proliferation and Safeguards* (Washington, D.C., 1977).

8. This account of the politics of the nuclear controversy in Sweden is based upon a series of interviews conducted by the authors in April, 1975 and May, 1976 with members of the Riksdag, government officials, and business leaders. We are especially grateful to Mr. Carl Tham, then Secretary of the Liberal Party, for his advice and assistance in arranging and carrying out these discussions. We are also deeply indebted to Mr. Mans Lonnroth, of the Secretariat for Future Studies, for his advice and assistance. See also: Mans Lonnroth, "Swedish Energy Policy: Technology in the Political Process," in *The Energy Syndrome*, Lindberg, ed., pp. 255–281.

9. Joseph B. Board, Jr., *The Government and Politics of Sweden* (Boston: Houghton Mifflin Co., 1970), Chapters 3 and 4.

10. Interview with Mr. Jungar Svonburg, Member of the Riksdag, Social Democratic Party.

11. Interview with Mr. Anders Wijkman, Member of the Riksdag, Conservative Party.

12. Soren Thunell and Lars Litjegren, "Energy Policy for Greater Prosperity," undated English translation of a report by the Social Democratic Party of Sweden, p. 3.

13. Interview with Mr. Jungar Svonburg, Member of the Riksdag, Social Democratic Party.

14. Interview with Mr. Carl Tham, Secretary of the Liberal Party.

15. Interview with Mr. Carl Tham.

16. Board, *The Government and Politics of Sweden.*

17. Interview with Mans Lonnroth, Secretariat for Future Studies.

18. Interview with Mr. Gunar Swenson, Member of the Riksdag, Communist Party.

19. Address to the meeting of the Executive Committee of the Social Democratic Labour Party, Sundsvall, February 1, 1975.

20. Ibid.

21. Interview with Mr. Anders Wijkman, Member of the Riksdag, Conservative Party.

22. Uno Svedin, "Sweden's Energy Debate," *Energy Policy* (September, 1975), pp. 258–261.

Chapter 9

1. Bupp et al., "Trends in Light Water Reactor Capital Costs in the U.S.: Causes and Consequences," Massachusetts Institute of Technology, Center for Policy Alternatives, (MIT: CPA 74–8), December, 1974, fig. 12. See also the statement of Harry O. Reinsch, Bechtel Power Corporation, before the Joint Committee on Atomic Energy, "Section 202" Hearings, February 6, 1974.

2. Bupp et al., "Trends in Light Water Reactor Capital Costs in the U.S.," fig. 13.

3. David L. Bodde, "Regulation and Technical Evolution: A Study of the Nuclear Steam Supply System and the Commercial Jet Engine," (Doctoral Thesis, Graduate School of Business Administration, Harvard University, 1975).

4. William E. Mooz, "Cost Estimating Relationships for Light Water Reactors," A Working Note prepared for the Energy Research and Development Administration, RAND Corp., WN–9926–ERDA, August, 1977.

5. Interviews with electric utility executives in France, Germany, and Sweden, June, 1976.

6. Bupp et al., "Trends in Light Water Reactor Capital Costs in the U.S.," fig. 16. This information, like the information referenced in the following paragraphs, was provided to the authors by American electric utility companies during the summer of 1974.

7. Ibid., fig. 17.

8. Ibid., Appendix III, p. 2.

9. Ibid., Appendix III, p. 10.

10. Interviews with executives of the American electric utility industry.

11. Natural Resources Defense Council v. Environmental Protection Agency, 478 Fed. 2d., 835 (1st Cir., 1973) (GERC 1879); National Resources Defense Council v. Environmental Protection Agency, 489 Fed. 2d., 390 (5th Cir., 1974) (GERC 1248); Sierra Club v. Ruckelshaus, 344 Fed. Supp. 253 (DDC, 1972).

12. Charles Komanoff, *Power Plant Performance* (New York: Council on Economic Priorities, 1976). An industry response to Komanoff's analysis of nuclear power plant performance can be found in: A. David Rossin, ed. "A Critique of the Report, *Power Plant Performance*," Commonwealth Edison Co., Chicago, January 27, 1977. In our judgment, this exchange leaves several questions open to further argument, but it shifts the burden of proof to the electric utilities and reactor manufacturers.

13. Spurgeon Keeney et al., *Nuclear Power: Issues and Choices* (Cambridge, Mass.: Ballinger Publishing Co., 1977):

"high estimate" $803/kw (1976 dollars)
"best estimate" $667/kw (1976 dollars)
"low estimate" $416/kw (1976 dollars)

Nuclear Engineering International, (November 1977):
$742/kw (1977 dollars)

Commonwealth Edison Co. (Telephone discussion with Alan Jakimo, research assistant, Harvard Business School Energy Research Project, July, 1977):

$534/kw (1976 dollars)

14. Spurgeon Keeney et al., *Nuclear Power: Issues and Choices:*

Ratio of Estimated Nuclear and Coal Plant Capital Costs

	For "high-sulfur" coal with pollution control equipment	*For low-sulfur coal without pollution control equipment*
"high estimate"	1.01	0.93
"best estimate"	0.83	0.70
"low estimate"	0.81	0.64

U.S. Energy Research and Development Administration, WASA-1345," (Washington, D.C., U.S. Government Printing Office, 1975):

	With pollution control equipment	*Without pollution control equipment*
600 MW plants	0.78	0.63
1,150 MW plants	0.82	0.67

15. Spurgeon Keeney et al., *Nuclear Power: Issues and Choices:*

	Fixed Charge Rates, Constant 1976 dollars	*Current dollars at 4 to 5 percent long-term inflation rate*
"high estimate"	15 percent	20 percent
"best estimate"	13 percent	
"low estimate"	10 percent	15 percent

Commonwealth Edison Co. (Telephone discussion with Alan Jakimo, research assistant, Harvard Business School Energy Research Project, July, 1977):

	Current dollars
"high estimate"	20 percent
"low estimate"	15 percent

Northeast Utilities Company (Telephone discussion with Alan Jakimo, research assistant, Harvard Business School Energy Research Project, July, 1977):

Current dollars
16.5 percent

16. Between 1964 and 1972, the fuel cycle cost component of nuclear power stayed between 1.5 and 2.0 mills/kwh (in current dollars). See: U.S. Energy Research and Development Administration, *WASH-1150*, (Washington, D.C.: U.S. Government Printing Office, 1969); and *WASH–1230* (Washington, D.C.: U.S. Government Printing Office, 1971). By 1974 this cost had risen to about 4 mills/kwh (also current dollars). See: U.S. Energy Research and Development Administration, *WASH–1345* (Washington, D.C.: U.S. Government Printing Office, 1975). By 1976, it was between 5.4 and 6.0 mills/kwh (in 1976 dollars). See: Spurgeon Kenney et al., *Nuclear Power: Issues and Choices*, p. 126. Confirmed by Commonwealth Edison Company by telephone conversation with Alan Jakimo, research assistant, Harvard Business School Energy Research Project, July, 1977.

The most thorough and lucid discussion we know of concerning the range of policy questions posed by the nuclear fuel cycle is: Atlantic Council of the United States, *Nuclear Fuels Policy*, Report of the Nuclear Fuels Policy Study Group of the Atlantic Council (New York, 1976).

17. L. Charles Hebel et al., *Report to the American Physical Society by the Study Group on Nuclear Fuel Cycles and Waste Management* (1977).

18. It is the standard practice of the American utility industry to express the anticipated costs of new generating plants in monetary units called "escalated dollars." This means that for each component of the plant, including labor, the company estimates the effect of inflation between the day of the estimate and the date when the equipment or service is actually paid for. For example, for any one kind of plant, the electric utility may know that a particular pump would cost $100,000 if it were to be delivered today. The utility must, therefore, estimate a future price for this piece of equipment. If, for example, inflation is predicted to be 6 percent per year for this item, and it is to be delivered one year hence, the purchase price will be estimated at $106,000. In developing an overall estimate of the cost of a power plant, $100,000 is included as the estimated "present-day" cost of the pump and $6,000 is included under the heading escalation. This process is repeated for all of the items—materials, equipment, and labor—in the total cost estimate.

There are two major sources of uncertainty here. The first is that the utility company has only an approximate idea of the exact date at which a particular item will actually be needed. The second is that the rate of inflation which will prevail until then is obviously uncertain. The result has been growing discrepancies among cost estimates as inflation rates have become unexpectedly steeper in the mid–1970s. The assumption of higher or lower future inflation rates is another way for either critics or advocates of nuclear power to manipulate the comparative economic outlook and thereby further bewilder or mislead the unwary.

It might seem to be an obvious question to ask why the electric utilities do not make their comparisons in "constant" dollars and thereby avoid this particular diffculty. In fact, reducing all cost estimates to constant dollars merely displaces the problem; it does not eliminate it. The reason for this has to do with the amortization of capital and it is subtle. Utilities call the average fraction of a capital expenditure that has to be depreciated each year of a plant's life, the "levelized annual fixed charge rate" for that plant. Although the value of this financial parameter is formally set as part of the regulatory process, it is in practice partly based on pragmatic estimates by utilities and rate-settling commissions of the cost of debt and equity financing. Such estimates necessarily take into account the perceptions of the financial risks associated with a planned capital investment. In a period of monetary inflation, interest rates demanded on debt are higher than they would otherwise be because they include some provision for the anticipated effects of inflation. Similarly, equity investors are generally more cautious in assuming the risks associated with long-term commitments. Both of these consequences of inflation are, therefore, implicitly contained in the fixed charge rate. It is extremely difficult to imagine precisely how such rates would differ in the absence of inflation. All of this is further clouded, if possible, by a theoretical disagreement among the experts. Many economists have called for the use of a "constant dollar" discount rate. For some economists the discount rate should reflect real returns in constant money and could be as low as 2 percent to 3 percent. For others, it should be determined exogenously by the government to reflect political choices about investment priorities. But because it is nearly impossible as a practical matter to predict the behavior of investors in the absence of inflation, this notion remains purely theoretical.

Whether the private sector should use the same discount rate as the public sector, even for equivalent investments (e.g., a uranium enrichment plant) is also an open question. In France, the government routinely establishes different discount rates for different sectors. (Varying from 10 percent to 12 percent for the Sixth Plan). The choice overtly reflects politically assigned economic growth priorities. This is why Electricité de France can routinely make investment comparisons in "constant francs" while an American utility would have great difficulty doing so. In the United States, therefore, the existence of monetary inflation further complicates the sort of ostensibly precise economic estimates that characterize much of the debate over the status of nuclear power.

A somewhat more technical demonstration of the wide range of coal prices within which the relative economic attractiveness of nuclear power is indeterminant is available from the Harvard Business School Energy Research Project, Soldiers Field, Boston, Mass., 02163. It is entitled "Working paper on the Relative Economics of Nuclear vs. Coal-Fired Generating Stations in the U.S."

Chapter 10

1. *Proceedings of the Colloque Sur les Implications Psychosociologiques du Développement de l'Energie Nucléaire,* Paris, January 13–15, 1977; published by the French Society for Radioprotection; see also: Francis Fagnani, *Le Débat Nucléaire en France* (Grenoble: ADISH, Paris-IREP, May, 1977); and: *Project RARE Opinion Survey Results,* August, 1977; Roger E. Kasperson, Graduate School of Geography,

Clark University, Worcester, Mass. 01610; also: Christoph Hohenemser, Roger Kasperson, and Robert Kates, "The Distrust of Nuclear Power," *Science* 196 (1 April 1977), pp. 25–34.

2. Louis Puiseux, *La Babel Nucléaire* (Paris: Editions Galilée, 1977), p. 170.

3. Spurgeon Keeney et al., *Nuclear Power: Issues and Choices* (Cambridge, Mass.: Ballinger Publishers, 1977).

4. Statement by the President on Nuclear Power Policy, April 7, 1977. For an insightful review and analysis of President Carter's policy, see: Michael Brenner, "Carter's Non-Proliferation Strategy: Fuel Assurances and Energy Security," Center for Arms Control and International Security Studies, Graduate School of Public and International Affairs, University of Pittsburgh, September, 1977. Cited by permission of the author. See also: Vince Taylor, "The Impossible Dream: The All-Electric, All-Nuclear, World Economy" (Paper presented at the Hearings on Nuclear Safeguards, Proliferation, and Alternate Fuel Cycles, Energy Resources Conservation and Development Commission, Sacramento, Calif., June 16, 1977). Finally, for background material on the Carter policy, see: U.S. Nuclear Regulatory Commission, *Final Generic Environmental Statement on the Use of Recycled Plutonium in Mixed Oxide Fuel in Light Water Cooled Reactors* (Washington, D.C.: August, 1976).

5. David Burnham, "Ford Discloses Nuclear-Curb Plan," *The New York Times*, 29 October 1976.

6. Spurgeon Keeney et al., *Nuclear Power: Issues and Choices* (Cambridge, Mass.: Ballinger, 1977), p. 3.

7. Ibid.

8. Ibid.

9. Interviews with French and German government officials, Summer, 1977. See also: Louis Puiseux, *La Babel Nucléaire.*

10. "Assessing the Nuclear Present and Future," *Nuclear News* 20 no. 10 (August, 1977), p. 33.

11. For a fuller and somewhat more technical discussion of the spent fuel/waste management problem, see: Alan Jakimo and Irvin C. Bupp, "Not in My Backyard," *Technology Review* (March/April, 1978).

12. Interviews with Carter Administration appointees to the Department of Energy.

13. David J. Rose, "The Dangers of Escalating Uncertainty," *Nuclear Engineering International* (September, 1977), pp. 66–67. See also: Alvin M. Weinberg, "Energy Policy and Energy Projections: The Case of a Nuclear Moratorium," (address to the Atomic Industrial Forum Conference on United States Energy Policy, Washington, D.C., January 11, 1977).

Chapter 11

1. The definitive history of Admiral Rickover's program is: Richard G. Hewlett and Francis Duncan, *Nuclear Navy, 1946–1962* (Chicago: University of Chicago Press, 1974).

2. Interview with Professor Lee Kowarsky, July, 1977.

3. Shaw believed that the light water manufacturers and their customers had seriously overextended themselves in jumping on the post-Oyster Creek "bandwagon." He was quite blunt on the point in meetings with the Chairman and members of the Atomic Energy Commission in early 1976. See: Bupp, "Priorities in Nuclear Technology," pp. 146–154.

4. For an account of the apparent lapses and omissions of the AEC's reactor safety program under Shaw see a series of articles by Robert Gillette: "Nuclear Safety: The Roots of Dissent," *Science* 177 (1972), p. 771; "The Years of Delay," p. 867; "Critics Charge Conflicts of Interest," p. 970; and "Barriers to Communication," p. 1030.

5. Beginning in 1957, the Joint Committee on Atomic Energy held annual public hearings on the "development, growth, and state of the atomic energy industry." The voluminous record of these hearings is an impressive monument to the mutually reinforcing government-industry capacity for self-deception.

6. The malfunctioning of the Joint Committee on Atomic Energy was apparent for at least a decade preceding its abolition in 1977. The definitive, and still pertinent account is: Harold P. Green and Alan Rosenthal, *Government of the Atom: The Integration of Powers* (New York: Atherton Press, 1963).

Chapter 12

1. These alternative conceptions have perhaps their most eloquent spokesman in Amory B. Lovins. See his: "Energy Strategy: The Road Not Taken," *Foreign Affairs* (October, 1976), pp. 65–97; and: *World Energy Strategies: Facts, Issues and Options* (New York: Friends of the Earth International, 1975).

2. Jean-Claude Derian and Andre Staropoli, *La Technologie Incontrolée* (Paris: Presses Universitaires de France, 1975); see also: Harvey Brooks, "Technology Assessment as a Process," *International Social Science Journal* 25, no. 3 (1973), pp. 247–256.

3. This observation is based on the authors' professional and personal acquaintances among members and supporters of the Union of Concerned Scientists, The Natural Resources Defense Council, and The Friends of the Earth.

4. One of the largest American environmental "lobbies," The Sierra Club, has often supported nuclear power in opposition to coal-fired plants.

5. F. Fagnani, "Le Débat Nucléaire en France: Acteurs Sociaux et Communication de Masse" Association pour le Développement de l'Information dans les Sciences de l'Hommes, Paris, May, 1977, p. 87.

6. *Proceedings of the Colloque Sur les Implications Psychosociologiques du Développement de l'Energie Nucléaire.*

7. Ibid.

INDEX